Problem-Solving in Mathematics

Critic of Institutions

Roberta Kevelson
Founder

William Pencak
General Editor

Vol. 16

PETER LANG
New York • Washington, D.C./Baltimore • Bern
Frankfurt am Main • Berlin • Brussels • Vienna • Oxford

Marcel Danesi

Problem-Solving in Mathematics

A Semiotic Perspective for Educators and Teachers

PETER LANG
New York • Washington, D.C./Baltimore • Bern
Frankfurt am Main • Berlin • Brussels • Vienna • Oxford

Library of Congress Cataloging-in-Publication Data

Danesi, Marcel.
Problem-solving in mathematics: a semiotic perspective
for educators and teachers / Marcel Danesi.
p. cm. — (Critic of institutions; v. 16)
Includes bibliographical references and index.
1. Problem solving—Study and teaching. 2. Mathematics—
Study and teaching. 3. Semiotics. I. Title.
QA63.D36 510—dc22 2008023832
ISBN 978-0-8204-5240-1
ISSN 1068-4689

Bibliographic information published by **Die Deutsche Bibliothek**.
Die Deutsche Bibliothek lists this publication in the "Deutsche
Nationalbibliografie"; detailed bibliographic data is available
on the Internet at http://dnb.ddb.de/.

29 Broadway, 18th floor, New York, NY 10006
www.peterlang.com

CONTENTS

PREFACE

Skill in math is becoming more and more of a practical necessity in today's information-based society. As journalist Thomas Friedman (2007: 300) has aptly put it, the reason for this is that the "world is moving into a new age of numbers" in which "partnerships between mathematicians and computer scientists are bulling into whole new domains of business and imposing efficiencies in math." Even the contemporary arts and entertainment worlds have taken notice. Popular television shows, such as the crime drama *Numbers*, and movies such as *A Hill on the Dark Side of the Moon, Good Will Hunting, Pi, Conceiving Ada,* and *Proof* (to mention just a few) are built around, or involve, mathematics as a basic element in their narrative makeup. But, despite its cinematic popularity, mathematics is seen by many as a struggle to learn and use. The Thinkport.org website, for instance, carried the following headline a few years back (Friedman 2007: 32): "How many times do we adults say to one another, I'm just not good at math?"

Given that the ability to count and understand basic arithmetical concepts (adding, taking away, etc.) is acquired effortlessly in childhood, why does this negative perception and even "fear" of math exist? Is it a consequence of how math is taught in school? Certainly the quality of the teaching materials used or the commitment of teachers is not at issue. Never before in the history of education have we had so many expertly-written math texts available for use as we have today and never before has the interest in teaching mathematics effectively been so fervent, as witnessed by the plethora of books, periodicals, websites, and the like devoted to its pedagogy. Nevertheless, many in the media (and society generally) still hold schools responsible for math phobia. Even celebrities have jumped onto this critical bandwagon. In her bestselling book titled *Math Doesn't Suck* (2007), television star Danica McKellar *(The Wonder Years* and *The West Wing)*, for example, suggests outright that fear of math starts in school.

Math phobia is a modern-day phenomenon. Common folk have always taken pleasure in numbers and math problems, not feared them. One of the oldest mathematical texts, an Egyptian work called the *Rhind Papyrus* written around 1650 BCE, is a collection of mathematical puzzles and problems composed (seemingly) for the pure joy of it (Peet 1923, Chase 1979, Danesi 2002). There has, in fact, always existed a "math recreation industry," involving some of history's most famous figures. Lewis Carroll (the author of *Alice's Adventures in Wonderland* and *Through the Looking Glass*), for instance, composed puzzles professionally for children and adults alike. Puzzle magazines, brainteaser sections in newspapers, not to mention crossword, sudoku, chess, and Rubik's Cube tournaments, appeal to hordes of ordinary people from every walk of life. Simply put, puzzles "bring us pleasure," as the late James Fixx (1983: 18) once wrote.

So, why is math perceived by so many to be so difficult and "unpleasant" today? If the schools are indeed to blame (which would of course be a sweeping oversimplification), then a fundamental reassessment of how math is defined and taught is required. In my view, any such reassessment must take into account the fact that math is, essentially, a *semiotic* discipline. This suggestion is not just an esoteric pet theory of mine. It is something that I have discovered practically. My goal in this book is, in fact, to show what a semiotically-based approach to the teaching of elementary math problem solving would entail. I have written it to share my experiences with teachers and educators, many of whom may not have come across the use of semiotics in math before. Since the book deals with basic math, I will exemplify and apply the relevant pedagogical principles primarily to elementary topics—numeration (chapter 1), arithmetic (chapter 2), algebra (chapter 3), story problems (chapter 4), geometry, combinatorics, and basic probability theory (chapter 5).

In the late 1990s, I designed and taught an experimental math course for "math phobics" at the University of Toronto. As a professor of semiotics (and an ex math major in university), I wanted to know for myself if the challenges posed by problem-solving in math could be tackled profitably by using materials designed semiotically. The experiment lasted 8 years, ending in 2006, producing positive outcomes, which I have documented and discussed in various journals (for example, Danesi 2003, 2007a). To my mind, that was the end of the matter. I would not have envisaged writing this book for a larger and more general audience, if it were not for an e-mail that a student sent to me in 2006. She was a self-declared math phobic who took my course "out of desperation" to pass a graduate school entrance exam. She wrote that she was finally able to understand the basic mathematical concepts for the first time in her life, thanks to the course, and hoped that others like her could gain the same confidence from the materials that I had used. This book is my answer to that student's entreaty.

The reader might find that many of the notions that I describe in this book are either common-sensical or perhaps "reformulations" of standard notions in math education. I suggest that this is so only in a coincidental way. My claim is that by perceiving math as a semiotic discipline, it will put the learner in a better intellectual position to acquire its basic structure effortlessly, akin to the way in which we learn other semiotic disciplines, such as language and music. I must also warn readers about what not to expect from this book. First, they will not find in it an in-depth treatment of math, problem-solving, or semiotic theory. My goal is simply to show how teaching problem-solving can (and should) be envisioned as a semiotic activity. Second, they should not expect to find a "prescription" of how to teach math. Semiotics is useful only as a heuristic tool, not as an all-encompassing method in education.

Finally, I should inform the reader that I am not a full-time instructor of mathematics, nor am I a professional mathematician. I am a semiotician who finds the most rewarding (and significant) uses of my discipline in the area of education. I continue to teach recreational mathematics at my university as an offshoot of this interest. I have become enormously enthusiastic about it, since it has produced some of the most gratifying outcomes of my professional career. It is something about which I now feel passionately. I cannot but agree with Mark Twain, when he wrote that: "Intellectual work is misnamed; it is a pleasure, a dissipation, and is its own highest reward" (quoted in *A Connecticut Yankee in King Arthur's Court*, chapter 28, 1889).

University of Toronto, 2008

1
PROBLEM-SOLVING

All science requires mathematics. The knowledge of mathematical things is almost innate in us. This is the easiest of sciences, a fact which is obvious in that no one's brain rejects it; for laymen and people who are utterly illiterate know how to count and reckon.

Roger Bacon (c. 1214–c. 1294)

The transformation of mathematics from a practical problem-solving craft to a theoretical discipline is traced generally to the method of proof perfected by Pythagoras (c. 582–500 BCE). He proved, for instance, that right triangles had the same abstract structure, no matter what the length of their sides (Maor 2007). The study of abstract structure is what mathematics is all about. So is semiotics.

Practical knowledge of the structure of right triangles is found in many parts of the world (from China to Africa and the Middle East) long before Pythagoras (Strohmeier and Westbrook 1999). Around 2000 BCE, the ancient Egyptians, for example, discovered that knotting and stretching a rope into sides of 3, 4, and 5 units in length produced a right triangle, with 5 the longest side (the hypotenuse). But it was Pythagoras who showed that this was really only one manifestation of a more general structural pattern—if the square of the length of the hypotenuse equals the sum of the squares of the lengths of the other two sides, then a right triangle is formed. Knotting any three stretches of rope according to this pattern—for example, 6, 8, and 10 units—will produce a right triangle because $6^2 + 8^2 = 10^2$ ($36 + 64 = 100$). As the historian of science, Jacob Bronowski (1973: 168) has insightfully written, we hardly recognize today how important Pythagoras' original demonstration (known as the Pythagorean Theorem ever since) was:

> The theorem of Pythagoras remains the most important single theorem in the whole of mathematics. That seems a bold and extraordinary thing to say, yet it is not extravagant; because what Pythagoras established is a fundamental characterization of the space in which we move, and it is the first time that it is translated into numbers. And the exact fit of the numbers describes the exact laws that bind the universe. If space had a different symmetry the theorem would not be true.

The goal of this chapter is to take a preliminary look at the role that semiotics can play in teaching problem solving (PS) and in examining mathematical structure generally. Among the first to recognize the importance of this role was the Russian psychologist Lev Vygotsky (1896–1934) whose work has become widely known in educational circles over the last few decades (Vygotsky 1962, 1978, Davydov and Radzikhovskii 1985, Daniels 1996). Vygotsky argued that true pedagogy should be based on connecting semiotics, the science of signs, psychological theory, the science investigating how signs are learned and remembered, and education, the instructional practices employed to teach individuals how to control signs.

Defining a Mathematical Problem

The archeological record shows that prehistoric people counted with their fingers, even before they had developed words for numerical concepts. Semiotically speaking, they used their fingers instinctively as *signs*. They also used other kinds of counting signs—pebbles, knots in a cord, marks on wood or stone, etc. From the latter the first number systems emerged. For instance, around 3000 BCE the Egyptians started using a decimal system based on counting groups of ten (without place value). Around 2100 BCE, the Babylonians developed a sexagesimal system based on counting groups of 60—a system we still use to this day to mark the passage of time. These early societies used their number systems primarily to solve practical problems—to survey fields, to carry out intricate calculations for constructing buildings, and so on. It was the Greeks who started analyzing the systems themselves, apart from their uses in solving practical problems. Around 300 BCE Euclid (c. 330–270 BCE), founded the first school of mathematics in Alexandria to study numbers, geometrical figures, and the method of proof. He also wrote the first true treatise of mathematical structure titled the *Elements*—a book that has permanently shaped how we teach mathematics. Euclid finished his proofs with QED, as it was later translated in Latin. The letters stood for *Quod erat demonstrandum* ("which was to be demonstrated")—remaining the symbolic hallmark of what mathematical method is all about to this day.

It was Plato (c. 427–347 BCE) who used the word *problem* to characterize a type of question, consisting of information or a set of conditions designed to elicit a particular answer (see Schoenfeld 1985, Lesgold 1988, and Barbeau 1992 for discussions of the term *problem*). Since antiquity, mathematicians have conceived of two basic types of problems—one in which the solution can be easily envisaged, and one in which it cannot. In English, the former is called a *problem* proper and the latter a *puzzle*. Generally speaking, the problem provides all the required information to reach a solution directly; the puzzle, on the other hand, provides information that appears to be incomplete or else that conceals a twist or a clever trap, thus making it much more difficult to reach a solution. Both types are used typically in classroom pedagogy.

The difference between the two can be illustrated with two problems from plane geometry:

[PROBLEM 1–1]
In the given figure, what is the value of x?

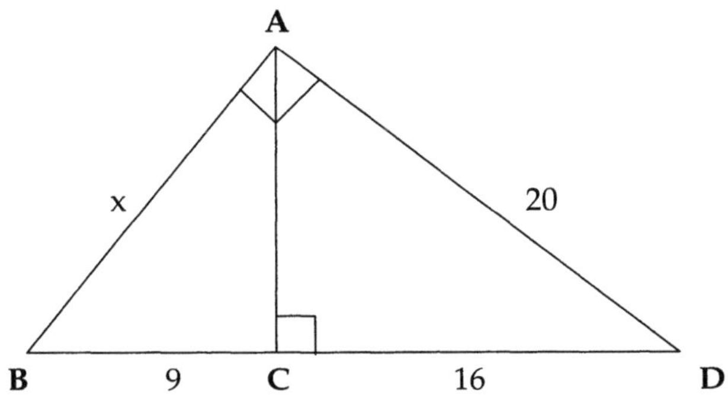

This problem has a straightforward solution because all the information that is needed to solve it is given to us (or else it can be easily fleshed out). ACB is a right triangle, and this suggests using the Pythagorean Theorem to figure out x, since x is the hypotenuse of that triangle. But first, we must figure out the length of AC (since it is not given directly). We notice that ACD is also a right triangle. So, to figure out the length of AC we use the Pythagorean Theorem:

$AD^2 = AC^2 + CD^2$	←	*Pythagorean Theorem*
$20^2 = AC^2 + 16^2$	←	*Plugging in the given values*
$20^2 - 16^2 = AC^2$	←	*Transposing 16^2 to the left side*
$AC^2 = 20^2 - 16^2$	←	*Turning the equation around*
$AC^2 = 400 - 256$	←	*$20^2 = 400$ and $16^2 = 256$*
$AC^2 = 144$	←	*Subtracting*
$AC = 12$	←	*Taking the square root of both sides*

Let's put the value of AC (= 12) on the diagram:

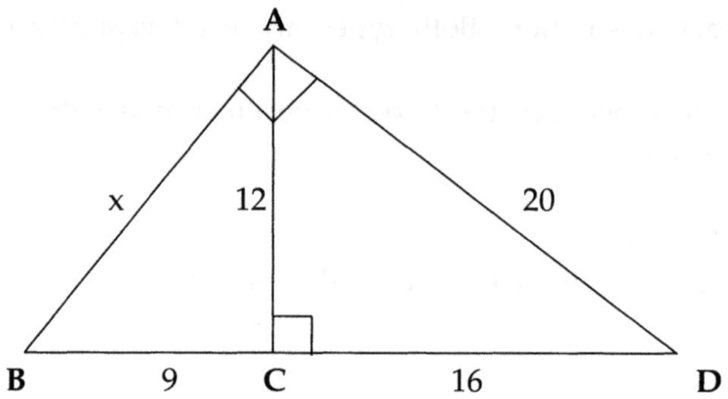

Now we can find the length of x easily:

$AB^2 = BC^2 + AC^2$	←	*Pythagorean Theorem*
$x^2 = 9^2 + 12^2$	←	*Plugging in the given values*
$x^2 = 81 + 144$	←	*$9^2 = 81$ and $12^2 = 144$*
$x^2 = 225$	←	*Adding*
$x = 15$	←	*Taking the square root of both sides*

We solved this problem easily by using the Pythagorean Theorem twice—a strategy suggested directly by the information contained in the problem. There are two things to note about the way the solution was presented that are central to the objectives of this book. First, a diagram was used (given to us in this case). Diagrams are semiotic tools, and good math teachers have always used them to teach all kinds of problems, not just geometric ones. A diagram is a *visual sign* (known more technically as an *icon*) that represents information concretely, and thus much more comprehensibly, by highlighting or showing what the essential features of the problem are, generally in outline form As widely-quoted math educator Richard Skemp (1971: 101) has put it, a diagram "conveys all that verbal statements do, more clearly and vividly." The pedagogical use of diagrams can thus be called *concretization.* Second, annotations were used to keep track of what was going on in the solution process. Each one is an *index,* semiotically speaking, which, like the indexes of books, allows us to refer to the relevant ideas, theorems, etc. involved and why they are being used. This pedagogical strategy can thus be called *annotation.* It serves as a guide, identifying in concrete terms what the stages and features of a solution are.

Now, consider the following problem, which at first glance would seem to suggest a similar use of the Pythagorean Theorem:

[PROBLEM 1–2]

Given the dimensions of the radius OD (6 + 4 = 10), can you calculate the length of the diagonal AB in rectangle AOBC?

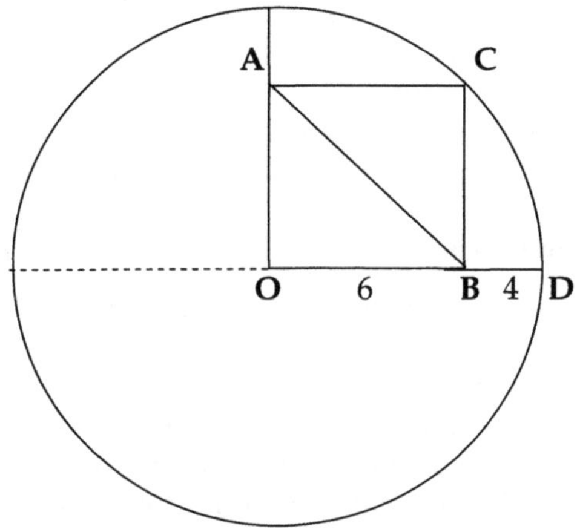

As it turns out, it is impossible to solve this problem by using the Pythagorean Theorem, as readers can confirm for themselves. There is simply not enough information on the given diagram to allow for a straightforward use of the theorem. So, let's consider what is known about circles and rectangles. First, the diagonals of a rectangle equal each other in length. This suggests drawing the other diagonal (OC) of rectangle AOBC:

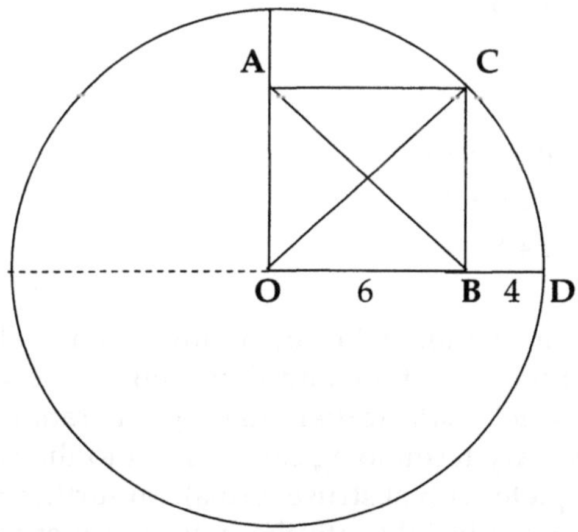

By doing this, we can see that diagonal OC is also a radius of the circle. The radii of a circle are equal. Line OBD is a radius and, as can be seen, is equal to 10 inches (6 + 4). So, line OC is also equal to 10 inches. From this we can conclude that the other diagonal, AB, is equal to 10 inches.

Students are typically fascinated by this problem. The likely reason is that it produces the same kind of surprise effect that a mystery story does. Using a distinction put forward by semiotician Umberto Eco (1979, 1992), between *open* and *closed* texts, [1–1] can be characterized as a *closed problem*, similar in how it presents information to such texts as recipes, guidebooks, etc.; and [1–2] can be characterized as an *open problem*, similar in its psychological effect and implicit construction to mystery stories.

Semiotics and Mathematics

Counting with fingers, or with substitutive signs such as pebbles and knots, is an instinctive semiotic act (as mentioned). These are called first-order signs. Numerals, on the other hand, are second-order signs—that is, they are signs derived from instinctive ones. We read second-order signs in terms of two simultaneous cognitive operations—differentiation and combination. The constituent digits in decimal numerals, for instance, take on specific values not only in terms of their actual physical shapes, but also in terms of the positions they occupy in the numeral. Thus, we read the values of 8 and 2 not only in terms of their differential forms (8 has a different shape than 2), but also in terms of their position in a numeral:

82
8 ← *Value* = 80
2 ← *Value* = 2
Total value: 80 + 2

28
2 ← *Value* = 20
8 ← *Value* = 8
Total value: 20 + 8

Determining the overall value of a numeral is called a *syntagmatic* (combinatory) cognitive operation, and determining the value of each digit in the numeral is a *paradigmatic* (differentiating) operation. While this might seem like an unnecessary terminological addition to the educational lexicon (which is already replete with abstruse terms), on further reflection it can be seen to provide a crucial insight into the nature of mathematical signs that may otherwise go unnoticed. In effect, the foregoing semiotic analysis allows us to understand how numerals are constructed in general and how we decode them. For instance, in the Roman numeral XIX (nineteen), determining the value of each sign is a paradigmatic operation: X = ten, IX = nine (one before ten). Determining the overall value of the numeral is achieved instead by combining the separate values: XIX = ten plus nine (nineteen).

The histories of semiotics and mathematics, when looked at comparatively, overlap ideationally and conceptually. (Readers interested in detailed treatments of semiotic history from various angles might consult Eco 1976, Deely 1982, 1990, 2001, Clarke 1987, Nöth 1990, Sebeok 2001, Petrilli and Ponzio 2005, or Danesi 2007b). The term *semiotics* (from Greek *sêmeiotikos* "observant of signs") was coined by the founder of western medicine, Hippocrates (c. 460–370 BCE), to designate the symptoms produced by the human body. Along with plant coloration, animal scents, and the like, symptoms are *natural signs* (produced by nature). On the other hand, signs produced by humans, such as numerals and words, are *conventional signs*. The latter are strategies for referring to and classifying the world. Among the first to point this out was Plato, who was intrigued by the fact that a word (verbal sign) such as *circle* does not refer to a singular thing, although it can if need be, as for example a circle with diameter of 2 inches, but to anything that has the property circularity, no matter what its diameter. Contrary to Plato (his teacher), Aristotle (384–322 BCE) argued that words start out as practical devices for naming singular things (circles with a specific diameter), rather than referring to generic properties (circularity, roundness, etc.). We discover that certain things have such properties as we go along naming them with our words. It is at such points of discovery that words start to stand for classes of things in general. From these we then go on to derive our categories.

It was St. Augustine (354–430 CE), the early Christian church father and philosopher, who first distinguished between natural and conventional signs in his *De doctrina Christiana*. For Augustine, natural signs *(signa naturalia)* included all natural phenomena, such as the bodily symptoms, the rustling of leaves, the colors of plants, the signals that animals emit, and so on. Conventional signs *(signa data)*, on the other hand, were the mental constructs created by people to serve their particular intellectual and social needs. These included not only words, but also gestures and visual symbols. Finally, St. Augustine considered miracles to be messages from God and, thus, sacred signs. These can only be understood on faith, although such understanding is partly based on specific religious interpretations of them.

Among the first to realize the value of understanding the function of signs in teaching and learning were the eleventh century educators known as the Scholastics. The Scholastics saw education as the process of acquiring signs to understand truths about the world. But among the Scholastics there were some—called nominalists—who argued that truth was a matter of subjective opinion and that signs captured, at best, only illusory and highly variable human versions of it. John Duns Scotus (c. 1266–1308) and William of Ockham (c. 1285-c. 1349), for instance, stressed that signs only referred to other signs, rather than to actual things—a perspective that is strikingly akin to some modern theories of the sign. St. Thomas Aquinas (1225–1274), arguably the greatest of the Scholastics, countered that signs referred to real things, but in variable ways. At about the same time, the English philosopher and scientist Roger Bacon (c. 1214–1292) developed one of the first comprehensive sign taxonomies, claiming that, without a firm understanding of the role of signs in the construction of knowledge, discussing what truth is or is not would end up being a trivial matter. A similar view was put forward by English philosopher John Locke (1632–1704) four centuries later.

The idea of fashioning an autonomous discipline of sign study did not crystallize until the late nineteenth century, when the Swiss linguist Ferdinand de Saussure (1857–1913) put such an idea forward in the *Cours de linguistique générale* (1916), a textbook compiled after his death by two of his university students. Saussure used the term *sémiologie* (English *semiology*), which he had used in personal correspondence as far back as 1894, to designate the new discipline. He also suggested that the main goal of semiology (should it ever come into being) was to understand the intellectual and social functions of signs.

The term *semiotics* is the preferred one today. It was adopted by the International Association of Semiotic Studies in 1969. The term *significs,* coined by Victoria Lady Welby (1837–1912) in 1896, is also used occasionally in the technical literature, but with a specific designation—the study of the relation of signs to the senses and the emotions. It was the American philosopher Charles S. Peirce (1839–1914) who put term *semiotics* (spelled originally *semeiotics*) into wide circulation (following Locke's suggestion in 1690). Along with Saussure, Peirce is a founder of modern semiotic theory and practice. Although his writing style is rather dense and his ideas not easily understood, Peirce's theory of the sign has become a central one. Perhaps his greatest insight is that signs are informed hunches as to what something means. Our experience of the world, therefore, influences how a sign is constituted and why it is brought into existence in the first place. Simply put, we create a sign not because we simply want to refer to something or classify it, but because we sense a relation between the sign and the thing it names.

Saussure claimed that we recognize something as a sign if it has a specific form and if it elicits a specific concept. He termed the form or physical structure of the sign, such as the sounds that make up the word *cat,* the *signifier,* and the concept that the sign elicits, the *signified* (literally "that which is signified by the sign"). The two, once linked, are perceived as inextricable psychologically. If one hears or reads the word *cat,* the image of the feline animal to which it refers comes instantly to mind; if one sees an exemplar of the feline mammal in question, and knows the word *cat,* then, analogously, this word is bound to come instantly to mind. Saussure claimed, moreover, that there is no intrinsic reason for creating the word *cat*, consisting of those specific sounds. Any other well-constructed signifier (word form) would have done the job just as effectively. This is why his model of the sign is called arbitrary. Peirce saw signs as possessing triadic structure—the actual physical sign, the thing to which it refers, and the interpretation that it elicits in real-world situations. He called the sign itself a *representamen* (literally "something that does the representing") and the concept that it encodes the *object* (literally "something cast outside for observation"). He termed the sense that we get from a sign the *interpretant*. This constitutes a derived sign itself, because it produces further meanings. In our culture, a *cat* is considered to be a domestic companion, among other things; in others it is viewed primarily as a sacred animal; and in others still it is considered to be a source of food. Thus, while the sign refers to virtually the same feline mammal in different cultures (no matter what name is used for it), its interpretant varies considerably.

Peirce also developed a comprehensive typology of signs. He identified 66 types in total. Newcomers to semiotics often react with perplexity to his typology, which consists of seemingly obscure and unfathomable notions. But the technical terms he coined are actually quite easy to figure out. For example, signs that draw attention to some quality of their referents (the objects they represent) are named *qualisigns*. In language, an adjective is such a sign, since it draws attention to the qualities (color, shape, size, etc.) of things. In other systems, qualisigns include colors (painting), harmonies and tones (music), etc. What the term qualisign allows to do, therefore, is to flesh out the common functions of signs across sign systems—a color in painting has the same function as an adjective in language, a tone or chord in music, and so on.

Despite the richness and breadth of Peircean sign theory, the Saussurean model continues to have wide usage among semioticians because it is a much more expedient one to apply, especially in the initial phases of analysis. Signifiers can easily be separated from contexts of occurrence and studied abstractly in relation to signifieds, albeit somewhat artificially. Peirce's model, however, has proven to be a more insightful and all-encompassing one in the development of a comprehensive theory of signs. It is accurate to say that semioticians today use a blend of Saussurean and Peircean concepts at various stages of analysis and for diverse purposes. They also frequently use ideas and findings from related or cognate disciplines, especially linguistics, philosophy, psychology, and anthropology. It should be noted, however, that this interdisciplinary mode of inquiry is a two-way street, since many ideas developed within semiotics proper are now found scattered in its cognate fields.

The historical link between semiotics and mathematics is an obvious one. In constructing appropriate signs (numerals, plus and minus signs, variables in algebra, equations, etc.) to represent numerical and spatial concepts, mathematicians have always been engaging in semiotic thinking. Since the late 1980s, this intrinsic link has caught the attention of math educators, as witnessed by the number of studies, special issues of journals, monograph series, websites, and a host of books dealing with the semiotics-mathematics interface, either directly or indirectly (for example, Cunningham 1987a, 1987b, 1992, 2007, Lemke 1987, Lave 1988, Rotman 1988, Varelas 1989, Reed 1994, Pimm 1995, Radford and Grenier 1996, MacNamara 1996, English 1997, Antenos-Conforti, Barbeau, and Danesi 1998, MacGregor and Price 1999, Lakoff and Nuñez 2000, Anderson, Sáenz-Ludlow, and Cifarelli 2000, 2003). As mentioned in the preface to this book, I believe that a semiotic approach to math education is useful as a means for putting learners in a position to understand the underlying sign-based structure of math and of how this can enhance PS skills.

Semiotics and Problem-Solving

Generally speaking, successful PS involves the ability to decipher correspondences between sets of signs (and what they stand for), and then representing them appropriately in order to grasp a solution strategy. This process is called simply *representation*. Consider a simple problem in cryptarithmetic—an open problem type that is solved by deciphering the correspondences that exist between letters and numbers (from Danesi 1997: 134–135):

> [PROBLEM 1–3]
>
> Decipher the following coded message, in which a specific number stands for a specific letter and given that 2 = S, 6 = P, 10 = B, and 12 = K. Each separate word is separated by a hyphen. The message relates to a type of book: 1 0 4 2—4 2—5—6 3 7 7 8 9—10 11 11 12.

The numbers and letters in this problem are signifiers. The message to be decoded is the signified. The process of replacing one type of signifier with another involves understanding how words and sentences are constructed in English. We start by replacing the numbers 2, 6, 10, 12 with the given letters: 2 = S, 6 = P, 10 = B, 12 = K. This will give us an initial glimpse of the hidden message:

> 1 0 4 S—4 S—5—P 3 7 7 8 9—B 11 11 K

Note that 5 is a single-letter word. The only two possible candidates for single-letter words in English are the pronoun *I* and the indefinite article *a*. Going through the English alphabet will reveal that it is impossible to form a word with the other letters used singly: *b, c, d, e,* and so on. The pronoun *I* usually starts a sentence: *I am going out; I love puzzles;* etc. There are, of course, sentences where this is not the case; *It is I who called; Maybe I should do it first;* etc. But it is always best to start out with the most likely possibility, which is the indefinite article *a*. So, by grammatical deduction let's replace 5 with A. We can always go back to this point in our signifier-replacement process if this deduction should not work out, and substitute 5 with the only other possibility (5 = I):

1 0 4 S—4 S—A—P 3 7 7 8 9—B 11 11 K

Now, let's focus our attention on the last word in the sentence: B 11 11 K. The sequence of two 11's means that the word contains two identical letters. These cannot be consonants, since no word consisting of four consecutive consonants exists in English (or any other language for that matter). So, we know that 11 11 stands for two identical vowels: *aa, ee, ii, oo,* or *uu*: The only replacement that produces a legitimate word is *oo*. The word is, therefore, 10 11 11 12 = B O O K. Our message now looks like this:

1 0 4 S—4 S—A—P 3 7 7 8 9—B O O K

Let's turn our attention to the two-letter word 4 S. Only a vowel is structurally viable as a replacement for 4 (otherwise we would have a word consisting of two consonants, which is impossible). So, the word could be either *as* or *is*. It is unlikely for *as* to occur as the second word in a five-word sentence. Composing a few five-word sentences will make this saliently obvious. So, once again, we should go with our most likely possibility, namely that 4 S = I S. Once again, should this turn out to be erroneous, we can always return to this point in our solution and go with our second option (4 S = A S). We replace 4 with I in the first word as well:

1 0 I S—I S—A—P 3 7 7 8 9—B O O K

Now, consider the first word. The only word that 1 0 I S can stand for is the demonstrative pronoun *This*. So, 1 = T and 0 = H:

T H I S—I S—A—P 3 7 7 8 9—B O O K

The remaining word, P 3 7 7 8 9, can now be considered in the context of the partially-deciphered sentence: *This is a p… book*. As can be seen, the word: (1) has six letters in it; (2) starts with the letter *p*; (3) has a sequence of two identical letters in it (7 7); and (4) refers to some kind of book. This suggests that *puzzle* is the required word, and, thus, that 3 = U, 7 = Z, 8 = L, and 9 = E:

T H I S—I S—A—P U Z Z L E—B O O K

The solution is: *This is a puzzle book.* Note, in summary, that the signifier-replacement process hinged on our knowledge of the structural characteristics of words and sentences. In other words, it revolved on knowing how linguistic signifiers are constructed and how they fit together in patterned ways. Pedagogically, it should be noted that, as in the case of problems [1–1] and [1–2], concretization was used (albeit in a different way) by displaying the partially-decoded verbal text after each signifier replacement; and annotation was used to explain how each signifier replacement was consistent with English word and sentence structure.

As the three examples of PS used so far illustrate, concretization and annotation may take on different forms, but they always involve, respectively, iconicity (representing something through resemblance or simulation) and indexicality (deducing something by relation or structural comparison). Consider now the following classic algebra problem:

> [PROBLEM 1–4]
> The other day my friend weighed her cat Fluffy on a balance scale. As it turned out, Fluffy weighed 20 pounds plus half of its own weight. How heavy was the cat?

We start by using a variable to represent the cat's unknown weight; that is, we start by replacing the weight with a letter-sign. It is customary to use the letter x in algebra. This type of sign is known more specifically as a symbol in semiotic theory. It is defined as a sign that has an arbitrary relation to its referent. Any other symbol (a dot, a line, etc.) could, in fact, have been chosen to represent the weight of the cat. The problem statement tells us that the cat weighed a total of 20 pounds plus half of its own weight. Consider the statement *half of its own weight*. Since the weight is represented by x, then half of this is represented by $\frac{1}{2}x$. Now, the problem states that the cat's total weight (x) was 20 pounds plus $\frac{1}{2}x$ (half its own weight). This can be represented by the following equation: $x = 20 + \frac{1}{2}x$. The symbols in the equation simply stand for the separate parts of the given statement. The equation translates the verbal text of the problem into a symbolic text. Solving the equation, we get $x = 40$. So, the cat weighed 40 pounds.

Teaching students how to solve this problem involves concretization and annotation. In this case, concretization entails breaking the problem down into its structural components using box diagrams. Below is one way in which this can be done:

X	=	20	+	$\frac{1}{2}$x
↓	↓	↓	↓	↓
cat's weight	*is*	*20 pounds*	*plus*	*half the total weight*

This type of diagram is called a *schema*. It is an iconic device that mirrors a balance scale in outline form, with x on the left-hand pan and 20 + $\frac{1}{2}$x on the right-hand pan. This means that the values on each side must be equal, otherwise the scale would not be balanced. The arrowheads on the schema are indexical devices showing what replacements are equivalent, providing visual annotation in this case (rather than verbal annotation as in the examples above).

At this point, it is instructive to point out that mathematicians have always been intuitively aware that mathematical PS is a semiotic process. They solved algebraic problems, at first, not by using letter symbols and the layout we call an equation, but geometric diagrammatic forms. For example, Euclid solved quadratic equations ($ax^2 + bx + c = 0$) by literally completing a square diagram. The so-called father of algebra, Diophantus (lived 3rd century CE), was the first to use symbols to stand for unknown quantities. It is no exaggeration to say that the history of mathematics unfolds fundamentally as a quest to find appropriate signs, symbols, and abbreviations to represent concepts, relations, or operations. The numerals, as mentioned, started out as finger signs. Here are a few other key "semiotic events," so to speak, that have marked mathematical history:

- Although the early Egyptians, Greeks, Hindus, and Arabs had symbols for addition and equality, the signs + and - were first used in 1489 by the German mathematician Johann Widman (1462–1498). It was the English mathematician Robert Recorde (c. 1510–1558) who devised the symbol = for equality in 1557.

- The English mathematician William Oughtred (1575–1660) first used the symbol × for "times" in 1631.

- The German mathematician Gottfried Wilhelm Leibniz (1646–1716) introduced the period to stand for multiplication. The Hindus wrote the divisor under the dividend in a ratio or division problem; Leibniz used the form a:b.

- The French mathematician René Descartes (1596–1650) was among the first to use the superscript notation a^n—a form that opened up, ipso facto, a new field of number theory (exponential numbers).

- The English mathematician John Wallis 1616–1703) defined the negative exponent and introduced the symbol ∞ for infinity in 1655.

- The English mathematician Thomas Harriot (1560–1621) originated the symbols > (greater than) and < (less than) in 1631.

- The Swiss mathematician Leonhard Euler (1707–1783) was responsible for the symbols Δ and f used in the theory of functions. He also introduced the use of the letter *i* for $\sqrt{-1}$ (the square root of minus 1) in 1748.

The late Thomas Sebeok (1920–2001) would often point out that the list of those who did semiotics without knowing it would fill the pages of an infinitude of books. He referred to this state of affairs as the "Monsieur Jourdain syndrome" in his lectures on semiotic theory. Monsieur Jourdain was a character in Molière's *Bourgeois Gentilhomme* (1670) who, when told that he spoke good prose, replied that he wasn't aware that it was prose that he had been using. Analogously, Sebeok would point out to a scholar or good teacher in mathematics who used semiotic notions unwittingly that he or she was, like Monsieur Jourdain, doing something of which she or he was not aware—semiotics.

Teaching Numbers as Signs

Numerals are signs that encode number concepts. They are so intrinsic to mathematical method that mathematicians rarely differentiate between the two in practice. This is akin to how we rarely make a distinction between words and the concepts they represent. We have, in effect, become habituated since childhood to perceive all signs and their meanings as *Gestalten* (signifying wholes).

The most basic are the counting numbers, known also as the *integers,* the *natural, cardinal,* or the *whole* numbers. These imply holistic quantitative concepts (rather than ones designating partiality), and are thus similar to the concepts encoded by mass nouns *(rice, information,* etc.). This is why the sum or product of whole numbers always produces another whole number:

$2 + 3 = 5$ ← *Two plus three equals (adds up to) five*

$2 \times 3 = 6$ ← *Two times three equals (produces) six*

On the other hand, dividing whole numbers does not produce another whole number, because division is akin to the process of partitioning something. As such, it is analogous to partitive structures in language (such as *some, little,* etc.): *She ate only some of the cake.* So, if we divide 2 by 3, we will not get another whole number. Rather, we will get a partitive number known appropriately as a *fraction*:

$\frac{2}{3}$ ← *Two partitioned three times = two thirds*

Various types of numerals (sign systems) are used to represent the same number concepts (referents). The numerical concept "two", for example, can be represented with the numeral 2 in the decimal number system, II in the Roman one, and 10 in the binary system. The connection between the numeral (or number sign) and its referent or concept, once established, is bidirectional or binary, as Saussure argued—that is, one implies the other. This is true of all signs (as discussed above). When we utter the word *tree* the image of an arboreal plant inevitably comes to mind and, in fact, we cannot block such an image from springing up in our brain; vice versa, when we see an arboreal plant the word *tree* seems to come also automatically to mind. Similarly, we have become so accustomed to using numerals that when the concept "two" comes up, we instinctively see the numeral 2 in our mind.

The ability to use numerals for number concepts seems not to be an innate skill, but rather an acquired one. The Swiss psychologist Jean Piaget demonstrated this with a set of ingenious experiments with children (summarized in Piaget 1952). For instance, when he showed a five-year-old child two matching sets of six eggs placed in six egg-cups, asking him or her whether there were as many eggs as egg-cups (or not), the child would invariably reply affirmatively. Piaget would then take the eggs out of the cups, bunching them together, with the egg-cups left where they were previously, asking the child whether or not all the eggs could be put into the cups, one in each cup and none left over. The answer he typically received was "no." Asked to count both eggs and cups, the child would correctly say that there was the same amount. But when asked if there were as many eggs as cups, the child would again answer "no." The child, Piaget concluded, had not grasped the relational properties of numeration, which are not affected by changes in the positions of objects. To the semiotician, Piaget showed, in effect, that five-year-old children have not yet established in their minds the binary connection between numerals and numbers. As the psychologist Skemp (1971: 154) has pointed out, Piaget's work has shown that counting is a learned skill that overlaps with verbal development: "counting is so much a part of the world around them that children learn to recite number-names not long after they learn to talk."

Numerals constitute a *code*—a historically-developed collection of signs with the same referential properties and structural features. The code we use commonly today is called *decimal* (from Latin *decem* "ten") because it is based on ten signs, called *digits* (meaning "fingers" in Latin). It is also called, logically, a *base-ten* system. The distinguishing feature of this code is that the value (actual magnitude) of any digit depends on the position it occupies in the numeral. This is, of course, a paradigmatic feature. As discussed above, the digit 2 has different values in the numbers 82 and 28, because it occurs in different positions in each. In teaching numbers to children, this feature of the decimal code appears to be one of the most difficult to impart conceptually, as the relevant literature (not to mention teaching experience) suggests. The difficulty seems to lie in grasping the fact that the value of a digit is determined by a rule—each digit in the numeral is 10 times greater than the value of the digit just to its right, and the total value that the numeral represents is determined by adding up the individual values of the digits that make it up. More technically, the value of a digit is determined by multiplying it by a successive power of ten (starting from the right side). This is, clearly, a syntagmatic property of the decimal code. The use of concretization and annotation techniques is thus absolutely critical in teaching how numerals are to be read. Take, for example, 248. Starting with the furthest digit to the right, it can be presented as follows:

8 ← *This has a unit value (8). Formally, it has the value:* 8×10^0 (8 *times* 10 *to the power of* 0).

4 ← *This has a value that is ten times greater* (4×10). *Formally, it has the value:* 4×10^1 (4 *times* 10 *to the power of* 1)

2 ← *This has a value that is ten times greater than the previous place and, thus, one hundred times greater than the first place to the right* ($2 \times 10 \times 10$). *Formally, it has the value:* 2×10^2 (2 *times* 10 *to the power of* 2)

Thus, 248 stands for "two hundreds + four tens + eight ones." As I have found through classroom experience, the directionality of presentation (starting from the right or from the left of the numeral) is irrelevant. Indeed, teaching to read numerals starting from both sides is a complementary and reinforcement technique. For example, 97,134 can be presented by starting from the left, this time—that is, with the digit that has the largest value:

9 = nine "ten thousands" ← $(9 \times 10 \times 10 \times 10 \times 10)$ or 9×10^4

7 = seven "thousands" ← $(7 \times 10 \times 10 \times 10)$ or 7×10^3
1 = one "hundreds" ← $(1 \times 10 \times 10)$ or 1×10^2
3 = three "tens" ← (3×10) or 3×10^1
4 = four "ones" ← (4×1) or 4×10^0

Again, the total value represented by the numeral is the sum of the individual values:

9 = ninety thousand ← 90,000
7 = seven thousand ← 7,000
1 = one hundreds ← 100
3 = thirty ← 30
4 = four ← 4
Total value: 90,000 + 7,000 + 100 + 30 + 4

The above presentation, as simple as it may seem, is nevertheless based on solid semiotic pedagogy—it is designed, in fact, to literally show what each digit stands for (concretization) and what this means in value terms (annotation). Now, for a place-value numeral code to work, a sign showing that a certain place can be left empty (without value) will be required. That symbol is, of course, 0. For example, to represent the quantity "ten," 0 is used in the "ones" place and 1 in the "tens" place. The result is 10. If we wish to represent the quantity "one hundred," we put 0 in both the "ones" place and the "tens place" and write 1 in the "hundreds" place. The result is 100. This can be presented as follows:

1 = one "hundreds" ← $(1 \times 10 \times 10)$ or 1×10^2
0 = no "tens" ← (0×10) or 0×10^1
0 = no "ones" ← (0×1) or 0×10^0

A practical example that would reinforce the teaching of 0 as a place marker is the car odometer—the device that tells us how far we have traveled. After traveling 9 miles, for example, the 9 on the odometer will be replaced by a 0 and the 1 digit will roll in to the left. When the car has traveled 99 (ninety-nine) miles, the 99 on the odometer will be replaced by 00 and the 1 digit will roll in to the left again, showing that we have traveled 100 (one hundred) miles. The use of practical examples from everyday life, which illustrate something, can be called *contextualization*. This can be defined simply as the pedagogical technique of providing real-world applications or analogues to illustrate some notion. The relevant research shows, in fact, that students who are given familiar, everyday contexts in which to locate math notions and problems increase their math skills more than do those who work on noncontextualized problems (Irwin 2001).

As an historical aside, it should be mentioned that the zero sign, and what it stands for, has been always a difficult one for philosophers and even mathematicians to accept. It was probably first conceptualized by the Babylonians, who left a blank space for it, and banned by the Greeks (of all people). The Chinese also left an empty space on their counting boards. There is archeological evidence that the Mayans had symbols for zero by about 250 CE and that the Hindus, who developed the symbol we use today, had devised it by the late 800s. The word *zero* derives from *ziphirum*, a Latinized form of the Arabic word *sifr* which, in turn, is a translation of the Hindu word *sunya* (void or empty). The Hindu zero symbol was introduced to Europe along with the decimal system by Leonardo Fibonacci (c. 1170–1240) in his 1202 book called the *Liber Abaci*. Fibonacci starts off his book, in fact, identifying zero as a numerical sign (cited in Posamentier and Lehmann 2007: 11):

> The nine Indian figures are: 9 8 7 6 5 4 3 2 1. With these nine figures, and with the sign 0, which the Arabs call zephyr, any number whatsoever is written.

A follow-up pedagogical technique for getting students to read numbers (numerals) correctly is, needless to say, *exemplification*. This means simply the process of constructing or adapting problems and activities for students that will allow them to utilize and apply the knowledge they have gained. The types of problems that can be constructed to exemplify the basic notions above are the ones that math manuals have traditionally used. Note, however, that they should be tailored to reflect the semiotic aspects of numerals. Here are a few prototypical examples followed by "semiotic commentaries," as they can be called:

[PROBLEM 1–5]
Write the following numbers as decimal numerals: "four thousand and fifty-five"; "six thousand, nine-hundred, and ninety-nine".

Semiotic Commentary: This type of problem is intended to get the student to grasp the difference between numerals as signs and number concepts (stated in words here).

[PROBLEM 1–6]

What is the value of 2 in the following numerals: 234, 2,340, 23,400, 234,000?

Semiotic Commentary: This type of problem is intended to get students to practice and reinforce the concept of positional value—a difficult characteristic of the decimal code.

[PROBLEM 1–7]

Indicate the value of each digit in the following numbers in terms of powers of ten: 94, 982, 4,561, 72,894

Semiotic Commentary: This type of problem is designed to get students to practice the concept of positional value by getting them to specify each digit value formally using exponent notation (10^n).

[PROBLEM 1–8]

Convert the following values into appropriate numerals: $2 \times 10^3 + 3 \times 10^2 + 1 \times 10^1 + 0 \times 10^0$; $9 \times 10^1 + 9 \times 10^0$

Semiotic Commentary: This type of problem is the reverse of [1–7].

As can be seen, these problems are closed texts that allow the student to reinforce the principles imparted in a controlled fashion. The use of open problems can also be used, and this will be discussed in due course. Suffice it to say here that, like learning to play a musical instrument proficiently, a certain amount of straightforward, quasi-mechanical (closed) practice is required.

A complementary semiotic technique that can be used to teach the decimal numbers is that of *opposition*. As is widely known, basic to Saussure's (1916) plan for the study of language (and other sign systems) was the notion of *différence* ("difference, opposition")—namely, the view that sign structures do not take on meaning and function in isolation, but rather in differential relation to each other. For example, the linguist determines the meaning and grammatical function of the word *cat* in English by opposing it to the word *rat* or some other minimal form. This opposition will show, among other things, that the initial consonants /k/ and /r/ are cues in English for establishing the differential meanings of the words. From such oppositions we can see, one or two features at a time, what makes the word *cat* unique in English, allowing us to pinpoint what *cat* means by virtue of how it is different from other words. At this microlevel of structure, opposition is clearly a technique that allows linguists to sift meaningful signals out from the phonic stream that constitutes the chain of speech. As a technique, opposition theory was elaborated by a number of linguists, many of Russian origin, who congregated in the Prague Circle in the early 1920s. The linguist Trubetzkoy (1968), for example, introduced the binary opposition technique of *minimal pair*—a pair of words, such as *cat* and *rat*, that differ by only one sound in the same position, which allows us to examine the difference in meaning between the two (as we saw). The minimal pair identifies the source of any *différence*, allowing the linguist to test empirically what features are significant in a language.

The Prague School linguists realized early on that extending the technique of opposition to levels of language beyond the phonic was a worthwhile endeavor. Opposition could be used, for example, to examine semantic relations such as synonymy (*big-large*), antonymy (*big-little*), taxonomy (*rose-flower*), part-whole relations (*handle-cup*), and so on. In effect, opposition provided a useful format for establishing the basic differential meanings of linguistic forms and structures. Opposition has rarely (if ever) been used as a technique in the teaching of math (as far as I know). Yet, like any sign system, numbers form sets of oppositions at different levels of representation and function.

The first critical opposition in the integer system is the one between *positive* and *negative* numbers. The positive ones are represented with the sign + in front:

$$\{+1, +2, +3, +4, \ldots\}$$

However, it is standard practice not to use this sign, unless it is needed for some specific purpose. This is so because such numbers are considered to be the unmarked numbers of the opposition—that is, the default form. The "unmarked" form of the opposition is, more technically, the representative (nonspecific) one, while the "marked" one is the exceptional form (in this case the negative numbers). For this reason it need not (literally) be marked with the + sign unless it is required for some specific reason:

$$\{+1, +2, +3, +4, \ldots\} = \{1, 2, 3, 4, \ldots\}$$

The negative integers, being marked forms, are always identified with the minus sign - in front:

$$\{-1, -2, -3, -4, \ldots\}$$

The complete set of integers (positive and negative) includes 0, which separates the positive and negative ones (as for example on a number line).

{..., -4, -3, -2, -2, 0, 1, 2, 3, 4, ...}

Negative numbers are used for a variety of practical reasons in daily life. So, at this point contextualization can be easily employed pedagogically to reinforce the concept of signed number. For example, any temperature marking "below" the zero point on a thermometer is indicated by a negative number. The temperature marking "5 below zero Centigrade" is represented as $-5\ C^{0}$. The student can thus be shown that the thermometer is really a number line—a line on which all the numbers, positive and negative, can be displayed with 0 in the middle as the "boundary point" between them. The negative numbers are placed to the left and the positive ones to the right of 0. Zero is not defined as either negative or positive. Again, it is a "placeholder" in the number code.

The distance between any two numbers is constant and, thus, the magnitude (value) of each number equals its distance on the line from zero:

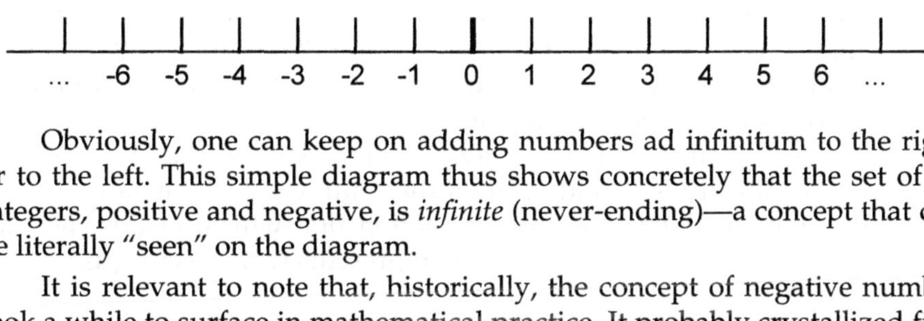

Obviously, one can keep on adding numbers ad infinitum to the right or to the left. This simple diagram thus shows concretely that the set of all integers, positive and negative, is *infinite* (never-ending)—a concept that can be literally "seen" on the diagram.

It is relevant to note that, historically, the concept of negative number took a while to surface in mathematical practice. It probably crystallized first in China, since the opposition of positive-versus-negative is found in a 250 BCE text titled *Chui-chang swan-shu* (The Nine Chapters). During the seventh century negative numbers are found in the bookkeeping practices and astronomical calculations of the Hindus. And it was not until the sixteenth century that such numbers surfaced in Europe, appearing in Girolamo Cardano's works. It is not coincidental that the term *negative* comes from the Latin past participle of *negare* ("to deny"), perhaps because the existence of such numbers had been denied for so long, or because it implied philosophically the "denial" of the positive, so to speak.

Another basic opposition that applies to the integers is that of *even* and *odd*—an opposition in which it is much more difficult to determine which of the two constitutes the marked and which the unmarked pole in it. The even pole is probably perceived as the unmarked one because its definition implies it—every even number is divisible by 2. An odd integer, on the other hand, is defined as any integer that leaves a remainder of 1 when divided by 2, thus constituting the derived (or exceptional) form in the opposition. This assessment also seems to have a practical validity, since we tend to count instinctively things in pairs. If we have 18 gloves, we can easily determine how many pairs we have by dividing 18 by 2, which gives 9 pairs. On the other hand, if we have 19 gloves (an odd number), we will have 9 pairs and a remainder of 1 glove when we divide 19 by 2. This means that in the latter collection of gloves there is, literally, an odd one in it. In effect, an even integer is a *multiple* of 2. A multiple is any number that is produced by the multiplication of smaller numbers, called *factors*. For example, the number 6 is a multiple of 3 and of 2 because it is produced when these two factors are multiplied together.

The question now becomes: Is there any way to represent the structure of even and odd numbers in general? In other words: What is their symbolic form? Any even number can be represented with the formula 2n where n can stand for any integer from the set of all integers:

$$n = \{\ldots, -4, -3, -2, -1, 0, 1, 2, 3, 4, \ldots\}$$

The use of letters to stand for numbers is, clearly, a symbolic strategy. The symbol 2n is an abbreviated version of $2 \times n$ (2 multiplied by n). It says that if we multiply any number by 2 the result will always be an even number. The process of representing a mathematical fact in this way, so as to reveal its structure, can be called, simply, *generalization*. It cannot be assumed, however, that the student understands this intuitively. So, it must be imparted through concretization and annotation first which, in this case, involves showing exactly what the formula represents with specific cases:

If $n = 1$ ← *Choosing* 1 *for* n
$2n = 2 \times n = 2 \times 1 = 2$
Result: an even integer

If $n = 2$ ← *Choosing* 2 *for* n
$2n = 2 \times n = 2 \times 2 = 4$
Result: an even integer

If $n = 3$ ← *Choosing* 3 *for* n
$2n = 2 \times n = 2 \times 3 = 6$
Result: an even integer

If $n = 4$ ← *Choosing* 4 *for* n

$2n = 2 \times n = 2 \times 4 = 8$

Result: an even integer

And so on.

The same result will emerge ad infinitum, no matter what integer value we assign to n, as the student can now easily see. The formula 2n is clearly a symbolic sign—standing for all even numbers. By plugging different values for n in it, the result is always an even number. In the above cases, we plugged in only positive values for n, producing positive even integers. But integers can be positive or negative, as discussed. To produce negative even integers, all we have to do is plug in negative integer values for n:

If $n = -1$ ← *Choosing* -1 *for* n

$2n = 2 \times n = 2 \times -1 = -2$

Result: an even negative integer

If $n = -2$ ← *Choosing* -2 *for* n

$2n = 2 \times n = 2 \times -2 = -4$

Result:→ an even negative integer

And so on.

As mentioned, when divided by 2, odd numbers leave a remainder of 1. If we consider the set of all the positive integers in sequence, as on a number line, we will see that a positive odd number is one more than (or one after) the preceding even number—3 is one more than 2, 5 is one more than 4, and so on. To put it another way, we can say that by adding 1 to 2 we get an odd number (3), by adding 1 to 4 we also get an odd number (5). For this reason, any odd number can be represented with the formula 2n + 1, which simply generalizes this observation. It is read as follows: by adding 1 to 2n (any even number), we will always get the number after it, which is odd. Again, using concretization with appropriate annotation, the meaning of the formula can be shown practically as follows:

If $n = 1$ ← *Choosing* 1 *for* n

$2n + 1 = (2 \times 1) + 1 = 2 + 1 = 3$

Result: an odd positive integer

If $n = 2$ ← *Choosing* 2 *for* n
$2n + 1 = (2 \times 2) + 1 = 4 + 1 = 5$
Result: an odd positive integer

If $n = 3$ ← *Choosing* 3 *for* n
$2n + 1 = (2 \times 3) + 1 = 6 + 1 = 7$
Result: an odd positive integer

If $n = 4$ ← *Choosing* 4 *for* n
$2n + 1 = (2 \times 4) + 1 = 8 + 1 = 9$
Result: an odd positive integer

And so on.

As above, the same result appears ad infinitum, no matter what integer value we assign to n. If we look at the number line, we will see that an odd number can also be defined as a number that is one less than (or one before) an even number: 1 is one less than 2, 3 is one less than 4, and so on. The formula that captures this observation is $2n - 1$:

If $n = 1$ ← *Choosing* 1 *for* n
$2n - 1 = (2 \times 1) - 1 = 2 - 1 = 1$
Result: an odd positive integer

If $n = 2$ ← *Choosing* 2 *for* n
$2n - 1 = (2 \times 2) - 1 = 4 - 1 = 3$
Result: an odd positive integer

If $n = 3$ ← *Choosing* 3 *for* n
$2n - 1 = (2 \times 3) - 1 = 6 - 1 = 5$
Result: an odd positive integer

If $n = 4$ ← *Choosing* 4 *for* n
$(2n - 1) = (2 \times 4) - 1 = 8 - 1 = 7$
Result: an odd positive integer

And so on.

So, to represent odd numbers in general we have two choices: 2n + 1 or 2n - 1. Needless to say, odd numbers can be negative. The same pedagogical technique can be used in this case too:

If n = -1 ← *Choosing -1 for* n
2n + 1 = (2 × -1) + 1 = -2 + 1 = -1
Result: an odd negative integer

If n = -2 ← *Choosing -2 for* n
2n + 1 = (2 × -2) + 1 = -4 + 1 = -3
Result: an odd negative integer

And so on.

Knowledge of how to represent numbers symbolically (with formulas) is extremely important in PS. It is pivotal, for example, in solving story problems involving digits, as we shall see in chapter 4. Problems such as the following ones can be used, at this stage, to get students to use symbolicity (the use of symbol formulas) practically, thus reinforcing it:

[PROBLEM 1–9]
What values of n in the formula 2n will generate all the even integers from 8 to 24 inclusive? For example: when n = 4, 2n = 8, because 2 × 4 = 8

Semiotic Commentary: This type of problem allows the student to use the general formula for even numbers to generate specific even numbers (within a given range) and, thus, to grasp the difference between the symbolic domain (the formula) and the referential domain (what it stands for).

[PROBLEM 1–10]
What values of n in the formula 2n + 1 will generate all the odd integers from 9 to 25 inclusive? For example: when n = 4, 2n + 1 = 9, because (2 × 4) + 1 = 8 + 1 = 9.

Semiotic Commentary: This problem has the same purpose as [1–9] for reinforcing odd number symbolicity.

[PROBLEM 1–11]
What value of n in either 2n (for even integers) or 2n + 1 (for odd integers) will generate each of the following integers: -5, -24, -31, -100?

Semiotic Commentary: This problem is a variant of the previous two.

[PROBLEM 1–12]
For each of the following questions determine whether the answer is an even or an odd integer. Use the general formulas for each type of number (2n for an even number and 2n + 1 for an odd number) as tools for assessing each case:

(1) What is the result of adding two even integers?
(2) What is the result of adding two odd integers?
(3) What is the result of adding an even and an odd integer?
(4) What is the result of multiplying two even integers?
(5) What is the result of multiplying two odd integers?
(6) What is the result of multiplying an even and an odd integer?

Semiotic Commentary: This problem allows the student to explore a given aspect of mathematical structure within the integer code. As such, it is an open problem.

Problem [1–12] requires further commentary. It constitutes a generic pedagogical strategy, which can be called simply *exploration*, rather than exemplification (which is the objective of the other problems in the above set). It can be defined as a strategy that allows the learner to use new information or skills in an exploratory fashion to see what they yield. The answer to (1) is even because $2n + 2n = 4n$ and 4n is itself even: 2(2n). The answer to (2) is also even because $(2n + 1) + (2n + 1) = 4n + 2$ and this is itself even: $2 (2n +1)$. The answer to (3) is odd because $2n + (2n + 1) = 4n + 1$ and this is itself odd: $(2 \times 2n) +1$. The answer to (4) is even because $(2n)(2n) = 4n^2$ and this is itself even: $2(2n^2)$. The answer to (5) is odd because $(2n + 1) (2n + 1) = 4n^2 + 4n + 1$ and this is itself odd: $2(2n^2 + 2n) + 1$. Finally, the answer to (6) is even because $(2n)(2n + 1) = 4n^2 +2n$ and this is itself even: $2(2n^2 + n)$. In all these problems the basic structure of even and odd numbers is examined and recast through substitutions and equivalences.

Another opposition that applies to the integers is that of *prime* versus *composite*. As mentioned, the factors of a number are the digits which, when multiplied together, produce that number. For example, 3 and 4 are factors of 12 because $3 \times 4 = 12$. The other factors of 12 are 1, 2, 6, and 12 itself. Clearly, some numbers can be decomposed into smaller numbers, called factors, which make them up through multiplication:

$$1 \times 12 = 12$$
$$2 \times 6 = 12$$
$$3 \times 4 = 12$$

In both semiotics and linguistics this type of analysis is called *morphological*. For example, in language, some words convey a "single piece of meaning." The word *logic* is such a word because it enfolds a single piece of meaning and cannot be decomposed further. On the other hand, the word *illogical* is composed of two forms, called *morphemes*, in addition to the minimal form *logic*. These are: the negative prefix *il-*, which conveys the meaning "opposite of;" and the suffix *-al*, which conveys the meaning "act or process of being something." Similarly, examining numbers such as 12 is akin to a morphological analysis.

Some factors can be decomposed further morphologically. For example, the factor 4 in the product $3 \times 4 = 12$ above can itself be expressed as the product of $2 \times 2 = 4$. We can plug this in as follows:

$$3 \times 4 = 3 \times (2 \times 2) = 3 \times 2 \times 2 = 12$$

This is also the case for $2 \times 6 = 12$, which can be decomposed to the same set of three factors, because 6 is itself composed of the factors $3 \times 2 = 6$:

$$2 \times 6 = 2 \times (3 \times 2) = 2 \times 3 \times 2 = 3 \times 2 \times 2 = 12$$

These are the skeletal factors of 12, since 3 and 2 cannot be decomposed further. This type of morphological analysis yields a fundamental insight into the structure of the integers and, thus, a central opposition within the integer code. The factors that cannot be decomposed further are called *prime numbers;* all the others are called *composite*. The definition of a prime is, in fact, any number that has no factors other than, of course, itself and the number 1 (which is a factor in every single number). For instance, the number 7 is prime because 7 and 1 are its only factors: $7 \times 1 = 7$. The number 1 is not defined as prime. The first eight primes are:

$$\{2, 3, 5, 7, 11, 13, 17, 19\}$$

Numbers are either prime or composite. The number 12, as we have seen, is a composite number because it is produced by the multiplication of prime factors, as we saw: $3 \times 2 \times 2 = 12$. It is obvious that the prime numbers are the marked ones, since there are many more composites and since they are themselves elemental numbers in the number code.

This prime-composite opposition is one of the most fascinating of all structural oppositions, having a long and fascinating history behind it. For example, as it turns out, each composite number is made up of one, and only one, distinct set of prime factors. This is called as the *Fundamental Theorem of Arithmetic*. It was formulated by Euclid (in Book IX of the *Elements*), and proved by the German mathematician Carl Friedrich Gauss (1777–1855) two millennia later. Given any composite number, such as 14 or 50, the theorem states that it is decomposable into a unique set of prime factors:

$$14 = 2 \times 7$$
$$50 = 2 \times 5 \times 5$$

Pedagogically, once students have been exposed to the prime-versus-composite opposition, it is a fairly simple task to demonstrate how to establish the unique set of prime factors that make up a composite number. Take 24 as a case-in point. The steps that can be used to reduce it to its skeletal prime morphology can be concretized and annotated as follows:

(1) $24 = 12 \times 2$
(2) Notice that $12 = 6 \times 2$
(3) Plug this in (1) above: $24 = (6 \times 2) \times 2 = 6 \times 2 \times 2$
(4) Notice that $6 = 3 \times 2$
(5) Plug this in (3) above: $24 = 6 \times 2 \times 2 = (3 \times 2) \times 2 \times 2 = 3 \times 2 \times 2 \times 2$

We have now uncovered the prime factors of 24. They are 2 and 3. Of course, the 2 occurs three times: $24 = 3 \times 2 \times 2 \times 2 = 3 \times 2^3$. We should remind the learner, at this point, that each of the prime factors in a composite number also divides evenly into it: 3 divides into 24 as does 2. This suggests a simple method for teaching the Fundamental Theorem (used typically in most textbooks). Take 220 as a case-in-point.

We can start by advising students to check if the smallest prime number, 2, divides into the number evenly:

$$220 \div 2 = 110$$

We tell students to continue dividing by 2 until it is no longer possible to do so (evenly):

$$110 \div 2 = 55$$

Now, 55 cannot be divided evenly by 2. So, we go to the next smallest prime in the set of primes, 3. It does not divide evenly into 55 either. Thus, we go on to the smallest prime after 3, which is 5. It does divide into 55 evenly:

$$55 \div 5 = 11$$

The result, 11, is itself a prime number, bringing the division procedure to an end. The prime factors of 220 are, therefore: 2 (twice), 5, and 11:

$$220 = 2 \times 2 \times 5 \times 11 = 2^2 \times 5 \times 11$$

It is a straightforward task to design problems that allow students to explore the properties of the prime-versus-composite opposition. Here are a few illustrative ones:

[PROBLEM 1–13]
Identify the following integers as either prime or composite:
5, 15, 51, 513, 659

Semiotic Commentary: This type of problem involves grasping the meaning of the prime-versus-composite opposition in a straightforward fashion.

[PROBLEM 1–14]
Determine the prime factors of the following composite integers: 45, 258, 5,688

Semiotic Commentary: This problem allows students to apply the algorithm discussed above for identifying the prime factors of a composite number.

[PROBLEM 1–15]

Can you find two primes that add up to the following even integers: 102, 200?

Semiotic Commentary: This is a more open type of problem that allows students to explore the nature of the relation between primes and composites.

I should mention at this point that using historical anecdotes and interesting aspects of a topic is a contextualization strategy. In my own courses, I have often observed that this renders the whole learning process much more lively and interesting. In the case of the primes, one could allude to the many fascinating problems and theoretical conundrums that their discovery has generated. Historical anecdotes intrigue and challenge students, providing a broader context in which to understand math concepts. A few of these are given below:

- Euclid showed that the number of primes is infinite (a proof that will be discussed in chapter 6). This proof seems to contradict common sense, since primes appear to become scarce as the numbers grow larger: twenty-five percent of the numbers between 1 and 100, 17 percent of the numbers between 1 and 1000, and 7 percent of the numbers between 1 and 1,000,000 are primes.

- Two primes that differ by 2, such as 5 and 7, are called twin primes. It is not known whether the number of twin primes is infinite.

- In the eighteenth century, Christian Goldbach (1690–1764) found that he could write every even integer greater than 2 as the sum of two primes: $4 = 2 + 2$, $6 = 3 + 3$, $8 = 3 + 5$, etc. So far, no one has been able to prove what has come to be known as Goldbach's Conjecture.

- To this day, no formula has ever been devised to generate primes and only primes. Marin Mersenne (1588–1648) thought he had come up with one such formula: $2^n + 1$. The formula does indeed generate a high number of primes, known today as Mersenne numbers. But many exceptions have been found. Pierre de Fermat (1601–1665) came up with a formula that he too thought would generate only prime numbers, but it did not.

- The mathematician Paul Erdös (1913–1996) proved that there is at least one prime number between any number greater than 1 and its double—a theorem know as Chebsyev's Theorem. For example, between 2 and its double 4 there is one prime, 3; between 11 and its double 22 there are three primes, 13, 17, and 19.

I have also found that commenting on the procedures used in a general way adds to the effectiveness of the whole pedagogical process. Such abstract commentary is, in effect, a type of contextualization strategy. For example, the procedures used above constitute an *algorithm* (a logical step-by-step procedure). An example of an early algorithm for the prime numbers is the so-called Sieve of Eratosthenes, invented in the 200s BCE by the Greek mathematician Eratosthenes:

- First write the series of whole numbers starting with 2:

{2, 3, 4, 5, 6, 7, 8, 9, 10, 11, 12, 13, 14, 15, 16, 17, 18....}

- Cross out every second number after 2, a procedure that eliminates all numbers that can be divided evenly by 2, except for 2 itself:

{2, 3, ~~4~~, 5, ~~6~~, 7, ~~8~~, 9, ~~10~~, 11, ~~12~~, 13, ~~14~~, 15, ~~16~~, 17, ~~18~~....}

- Similarly, cross out every third number after 3, a step that eliminates all the numbers that can be divided evenly by 3, except for 3 itself:

{2, 3, ~~4~~, 5, ~~6~~, 7, ~~8~~, ~~9~~, ~~10~~, 11, ~~12~~, 13, ~~14~~, ~~15~~, ~~16~~, 17, ~~18~~....}

- Then, since 4 is already crossed out, move on to 5. Continue the same process. Any number that has not been crossed out is prime. They can be thought of as having passed through a sieve (strainer) that has removed all the composites.

This shows what an algorithm is and allows one to do. Incidentally, the word comes from the name of the twelfth-century Baghdad scholar Muhammad ibn Musa al-Khwarizmi, whose treatise *On the Hindu Art of Reckon-*

ing was translated into Latin, as was his name (Algoritmi). Contextualization strategies are absolutely critical in reinforcing learning. These vary widely, from using everyday applications of some math principle to historical anecdotes.

The final opposition to be discussed here is the *integer*-versus-*fraction* one, in which the *fraction* is, clearly, the unmarked form of the opposition. A concretization technique used in math pedagogy is that of the pie (or some other figure or common item). Students can then be told that the opposition encodes a whole-part system of reference. So, in the case of the integers, a digit such as 1 represents a whole pie, a digit such as 2 stands for two whole pies, and so on. To represent a part of some entity, such as one half or one-third of a pie, we need another type of numeral, known as fraction, in which two digits—one over the other—are used to indicate the type of partitioning involved.

$\frac{1}{2}$ ← *One half*

$\frac{1}{3}$ ← *One-third*

The digit on top is called the *numerator* and the one on the bottom the *denominator*. The sign is iconic (imitative of its referent) and this can be imparted by pointing out that if one cuts a pie equally in half, one will get two equal pieces. Each of these is "one of two" or $\frac{1}{2}$ (also written as 1/2). Now, if we cut the same pie into three equal parts, we will produce three equal pieces. Each of these is "one of three" or $\frac{1}{3}$ (also written as 1/3). Appropriate diagrams are the following:

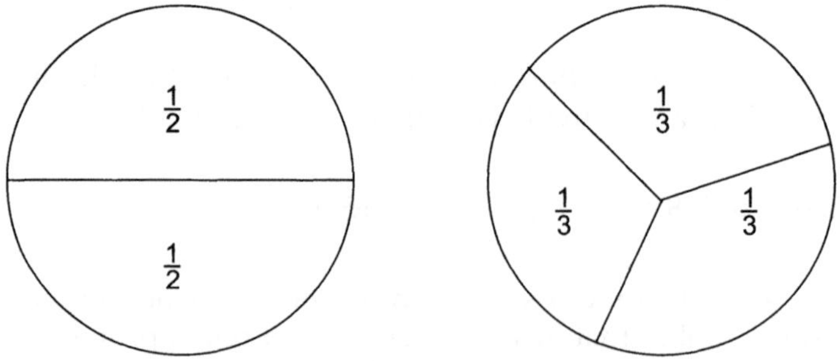

Now, the student can be asked to compare the value of the two fractions. As can be seen from the diagrams, 1/2 represents a larger piece of the pie than does 1/3. Thus, the fraction 1/2 is greater in value than 1/3. This is formalized as:

$1/2 > 1/3$

It is critical to get students to grasp this conceptually, for as research shows, many have a hard time believing that a fraction such as 1/3 is smaller than 1/2 because the denominator digit is larger in the former (Post, Wachsmuth, Lesh, and Behr 1985). Semiotically speaking, students are misreading the sign-form and teachers must thus focus their minds on what the sign stands for. As the diagrammatic technique shows, if the numerator of two fractions is the same, the one with the lesser denominator has a larger value.

Now, if we compare the values of fractions with different numerators but identical denominators, such as 3/4 and 2/4, and use appropriate diagrams, we can show that the numerator in this case indicates the larger value. Both fractions can be envisioned as indicating that a pie was cut into four equal pieces. The numerator of 3/4 tells us that the fraction stands for "three" of those pieces, whereas the numerator of 2/4 tells us that it stands for only two of the same pieces. Obviously, three is greater than two, so the value of the fraction 3/4 is greater than that of 2/4.

Concretizing fractions with different numerators and different denominators is a little trickier. Consider 2/3 and 5/9. First, we can tell students to look at their denominators. The denominator of 2/3 tells us that a pie has been cut into three equal pieces and the denominator of 5/9 tells us that it has been cut into nine equal pieces. The fraction 2/3 stands for "two of the three" pieces; and the fraction 5/9 stands for "five of nine" pieces. Can these be compared? First, three of the pieces of the 5/9 pie are equal to one of the 2/3 pie. A pie diagram will show this concretely. Thus, six of the nine pieces, or 6/9, would be equivalent in value (actual amount of pie) to two of the three pieces:

$6/9 = 2/3$

This is why these are called *equivalent* fractions. Now, 6/9 is greater than 5/9 because it has a greater numerator (recalling the principle above). Thus, 2/3 (which is equivalent to 6/9) is also greater than 5/9. Let's lay out the reasoning in sequence, using annotations as follows:

(1) $6/9 > 5/9$	← *Larger numerator, same denominator*
(2) $6/9 = 2/3$	← *Equivalent fractions*
(3) $2/3 > 5/9$	← *Plugging* 2/3 *in* (1) *in place of* 6/9

At this point a shortcut for producing two equivalent fractions can be introduced: multiply one or both by factors that produce a common denominator. In the case of the fraction 2/3 above, we simply multiply the numerator and the denominator by 3:

$$\frac{2 \times 3}{3 \times 3} = \frac{6}{9}$$

The 9 is the common denominator. And we remind the student that if we multiply the denominator by 3 we must also multiply the numerator by 3 (in order to keep the value of the fraction the same).

At this point, a generalization strategy can be used to summarize the integer-versus-fraction opposition. A fraction is really a *ratio* of two integers, one on top of the other, as the student has by now grasped. If p stands for any numerator and q for any denominator, a fraction can be represented with the general formula p/q, as long as $q \neq 0$ (read "q does not equal 0"), because division by 0 is not allowed in mathematics. Actually, an integer can also be represented with the same formula. In this particular case q has a constant value: namely, 1. So, for example, the first positive integer 1 can be represented as 1/1, the second one 2 as 2/1, and so on. The set of integers can now be shown to have the same structure of the fractions:

$$p/q = \{1/1, 2/1, 3/1, 4/1, \ldots\} = \{1, 2, 3, 4, \ldots\}$$

Because of this structural feature, revealed by using the same sign form, both the integers and the fractions are classified together and called *rational numbers* (literally, numbers that can be represented as ratios). Any number that cannot be represented with a p/q form is called *irrational*. One such numbers is $\sqrt{2}$. Taking the square root of $\sqrt{16}$ produces an integer, namely 4. However, if we try to take the square root of $\sqrt{2}$, we will get a decimal with numbers after the point repeating at random:

$$\sqrt{2} = 1.4142\ldots$$

At this point, the teacher can complete the pedagogical process by indicating that fractions can be expressed equivalently in so-called decimal form. In this case a decimal point is used in the make-up of the numeral sign. Each place to the right of the point decreases in value by one tenth. So, the first one to the right has a 1/10 fractional value; the second place to the right the value of 1/100; and so on. Take, for example 82.537. If we multiply $5 \times 1/10$

we will get 5/10, which is written in decimal form as .5. If we multiply 3 × 1/100 we will get 3/100, which is written in decimal form as .03. And, if we multiply 7 × 1/1000 we will get 7/1000, which is written in decimal form as .007.

A fraction form can be converted to equivalent decimal form by dividing the denominator into the numerator. In some cases the resulting decimal is a repeating one:

$$1/3 = 1 \div 3 = 0.33333...$$
$$8/9 = 8 \div 9 = 0.88888...$$

We can also point out that 0 is used to the left of the decimal point to indicate that there are no digits (with integer value) before the period in the numeral. In the conversion of other fractions to decimals, a so-called terminating decimal might result instead:

$$1/4 = 1 \div 4 = 0.25$$
$$2/5 = 2 \div 5 = 0.4$$

Some conversions produce repeating sequences of digits. For example, converting the fraction 3/7 produces the repeating six-digit sequence 428571 ad infinitum:

$$3/7 = 3 \div 7 = 0.428571\ 428571\ 428571\ 428571...$$

This is called, logically, a *periodic* number. Incidentally, as mentioned the 0 tells us that there are no whole digits in the converted decimal numeral, and that this means that the numerator is smaller than the denominator in the original fraction. Fractions in which this is the case are called *proper*. An *improper* fraction, on the other hand, is one in which the numerator is larger than the denominator, such as 6/5, 9/2, and so on. Improper fractions can be converted to *mixed fractions* by dividing the denominator into the numerator and expressing any remainder as a fraction. Thus:

6/5 = 6 ÷ 5 = 1 and 1/5 (remainder) → *Written* 1 1/5
9/2 = 9 ÷ 2 = 4 and 1/2 (remainder) → *Written* 4 1/2

And, of course, these can be converted to mixed decimals as shown:

1 1/5 =	1 + (1 ÷ 5)	=	1 + .2	=	1.2
4 1/2 =	4 + (1 ÷ 2)	=	4 + .5	=	1.5

Finally, we can introduce the concept of the *absolute value* of any number n, shown as $|n|$. This is the value of the number regardless of its sign.

Number	*Absolute Value*		
0	$\|0\|$	=	0
15	$\|15\|$	=	15
21	$\|21\|$	=	21
1/2	$\|1/2\|$	=	1/2
-15	$\|-15\|$	=	15
-199	$\|-199\|$	=	199
-4/5	$\|-4/5\|$	=	4/5

And so on.

The absolute value tells us what a number's actual distance from 0 is, without regard as to whether it is to the left or to the right of 0 on the number line. So, $|5|$ and $|-5|$ have the same absolute value (5), and this means simply that they are equidistant from 0. Obviously, 5 lies to the right of 0 and -5 to the left.

Problems related to the practice of the sets of oppositions discussed above can now be easily constructed, involving various degrees of openness and closure. A few are as follows:

[PROBLEM 1–16]

Place these numbers on the number line: $\sqrt{2}$, -1/2, $-\sqrt{3}$, 5.3, 7/8

Semiotic Commentary: This type of problem allows the students to visualize concretely what each numerical sign represents.

[PROBLEM 1–17]

Give the absolute value of the following: $|4|$, $|-4|$, $|3/4|$, $|-3/4|$, $|\sqrt{2}|$, $|-\sqrt{2}|$, $|4.056|$, $|-4.056|$

Semiotic Commentary: This allows students to concretize the difference between value and directionality (expressed by the positive-versus-negative opposition).

[PROBLEM 1–18]

Indicate what type of number each one is as completely as possible: 22, -23, $\sqrt{3}$, 3/4, 0.454545, -3.45, -7/5

Semiotic Commentary: This simple recognition exercise is still a valuable one since it gets students to connect the signifiers used for numbers and what these mean in mathematical terms.

Math Competence

In this chapter I have argued, implicitly, that math competence is related to semiotic competence—the ability to understand sign structures and how they represent things (iconically, indexically, symbolically, etc.). Semiotic pedagogy involves imparting this ability through techniques derived from semiotic practices and concepts (opposition, concretization, annotation, generalization, contextualization, analogies to real life, etc.). These techniques are summarized in the glossary at the back of this book. In the area of PS, math competence can be defined therefore as the ability to represent a problem in an appropriate semiotic fashion, whereby the problem is converted into an appropriate sign-form (known more technically as a *representamen*).

Consider an engineering problem. Suppose that a tunnel has to be dug right through the middle of a boulder. The length of the tunnel cannot obviously be measured directly. How do we solve this problem? An easy plan of how to do so is made possible by the Pythagorean Theorem. A point A on one side of the boulder and another point B on the other are chosen such that both points remain visible from a point C to the right, with C chosen so that angle ACB is a right angle (90^0). Then, by aligning A with A′ (the entrance to the tunnel on one side) and B with B′ (the entrance to the tunnel on the other side) the required length can be determined. First, it is a straightforward task to measure AC and BC, since these lengths have been chosen on purpose to permit their measurability. Second, the Pythagorean Theorem tells us that $AB^2 = AC^2 + BC^2$. This formula allows us to discover the length of AB easily because the lengths AC and BC are known. By subtracting the distances AA′ and BB′ from AB, the length of the tunnel (A′B′) can now be determined. It is: A′B′ = AB - (AA′ + BB′),

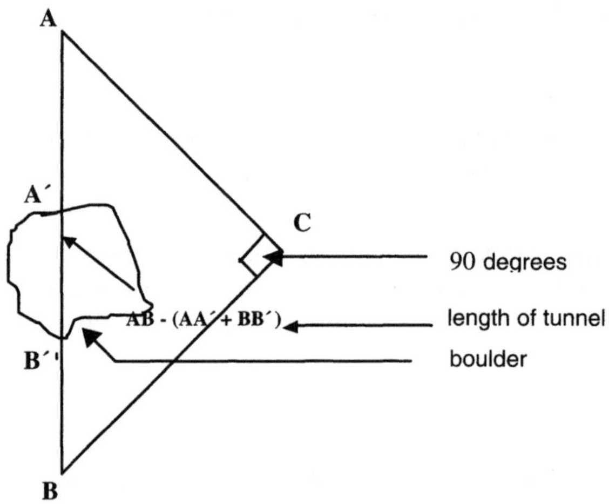

The above diagram is, as discussed throughout this chapter, an iconic representamen, showing in outline form the configuration of the actual physical scenario described in the problem. In the history of science, this kind of reasoning has produced truly remarkable results. This is because, as the ancient Greeks emphasized, it permits representations of various aspects of physical reality which, in turn, allow for an intellectual experimentation with that very reality through the use of symbols and symbolic reasoning. The results of this experimentation can then be redirected to the real world to see what they yield.

The pedagogical notions introduced in this chapter are not intended simply to replace other ones in current pedagogical practice. Cumulatively, they suggest a sequence of classroom procedures and stages that, arguably, reflect more accurately the psychology of learning math and PS. To test this hypothesis, I conducted the following simple pedagogical experiment a few years ago. I separated the students in the "math for math phobics" course I used to teach into two sections—each with 6 students (3 females and 2 males in each section). I could not, of course, control for the usual variables such as age, educational background, and the like, since only those that had enrolled in the course could be chosen. The sample was, however, random in the basic sense of that word. To avoid any experimenter bias, I asked two research assistants to teach the two sections. Section A was taught using the semiotic principles discussed in this chapter and section B with another (more "traditional") type of pedagogy. The assistant teaching A was trained in advance to use the semiotic techniques discussed here; the assistant teaching B (who was a math major) was allowed to teach in any way he thought was appropriate, given his own teaching experiences. The same textbook and course materials were used in both sections.

At the end of the course a standardized test (similar to the one used on the GMAT) was given to both groups. The results were as follows. Section A scored (on average) 86% correctly; section B scored 62% correctly. Needless to say, the experiment was informal and lacked the usual rigor. But it did bring out nonetheless that semiotic pedagogy could potentially make a difference.

In my view, the semiotic framework discussed here (and in the remainder of this book) will provide insights on how learning and representational processes overlap and intertwine and these can be converted into teaching practices and principles. Novel learning tasks are more apt to be more comprehensible if they are presented iconically (through concretization, annotation, and contextualization); whereas successful practice of already-acquired knowledge and skill suggests the use of techniques such as exploration and generalization. The key to successful learning lies, arguably, in determining at what point the student's learning mode shifts from concretization to generalization. Actually, as Vygotsky (1962) pointed out, learners will demonstrate by their behavior when they have reached that point, which he called the next or proximal "zone" of development. The zones will, of course, vary from student to student in terms of breadth (how much a particular individual is capable of learning something) and duration (the length of time the individual needs to grasp some learning task); but the attainment of each one will show up invariably, for a particular student, in identifiable representational behavior.

2
PROBLEMS IN ARITHMETIC

I know that two and two make four and should be glad to prove it too if I could—though I must say if by any sort of process I could convert 2 & 2 into five it would give me much greater pleasure.

Lord Byron (1788–1824)

Arithmetic is the art of making and interpreting numerical models (signs) of common sense counting notions. We instinctively use this "count modeling" art when, for example, we determine how many things in total are represented by the number of fingers raised by someone on two hands, when we compare the number of toes to the number of fingers we have, and when we conclude that if we cut two fingernails on one hand we end up having three longer fingernails on that hand. The word *arithmetic* comes from the Greek word *arithmetike*, which means, literally, "the art of counting." All the ancient civilizations practiced this art, using it in business transactions, in measurement, and in many other practical social activities long before they had alphabets. When alphabets appeared on the scene around 1000 BCE, they were, as a matter of fact, also used to represent the numbers. The order {A, B, C, …} of our own alphabet is based on that early practice, where A stood for the number 1, B for the number 2, and so on. It was the ancient Greeks who started the study of arithmetic as a logical system, separate from its practical uses, beginning in the 500s BCE with Thales (c. 625–546 BCE) and Pythagoras.

The branch of mathematics that deals with the theoretical foundations of arithmetic is now known as *number theory*. It was the French mathematician Pierre de Fermat (1601–1665) who proposed that the goal of number theory be to explore the structure of the arithmetical forms and operations. A landmark book in number theory, *Disquisitiones Arithmeticae* (1801), was written a century and a half later by the German mathematician Carl Friedrich Gauss (1777–1855). In it, Gauss developed a comprehensive theory of natural numbers, including a proof of the Fundamental Theorem of Arithmetic (discussed in the previous chapter). In semiotic terms, number theory is really sign theory, since the arithmetical operations (addition, subtraction, etc.) constitute a code. Given this basic fact, arithmetical PS can be taught with the same kinds of principles, strategies, and techniques as those introduced in chapter 1.

Defining an Arithmetical Problem

A problem in arithmetic can be reduced conceptually to a problem in counting. Following is a simple problem that encapsulates what an arithmetical problem is essentially all about (from Danesi 2006a: 23):

> [PROBLEM 2–1]
> Debbie has $.25 cents in her left pocket and $1.35 in her right. Her friend Wendy has $.45 cents less than Debbie has altogether, while another friend, Monica, has $.55 cents more than Wendy has. The girls want to buy a CD that is on sale for only $5.00. Do they have enough money to buy it, if they pool all their financial resources?

This problem can be broken down into its arithmetical components as follows. We are told that Debbie has money in two pockets—$.25 in the left pocket and $1.35 in the right one. This implies addition. So, altogether Debbie has:

$.25 + $ 1.35 = $1.60

Now, how much money does Wendy have? Wendy has $.45 cents less than Debbie has altogether. So, to find out how much money she has, all we have to do is "take away" $.45 from Debbie's amount. This implies subtraction:

$1.60 - $.45 = $1.15

That's how much money Wendy has. It is, of course, $.45 cents less than Debbie has. How much money does Monica have? It is easy to figure out how much money she has, since we are told that she has $.55 more than Wendy. How much money does Wendy have? As we have just found out, it is $1.15. What is $.55 more than $1.15? It is shown with the following addition representamen:

$1.15 + $.55 = $1.70

What do we know now? We know how much money each girl has: Debbie has $1.60; Wendy has $1.15; and Monica has $1.70. How much money do the girls have all told?

Debbie has	\$1.60
Wendy has	\$1.15
Monica has	\$1.70
Total	\$4.45

Can they afford to buy the CD? The answer is no, since the CD costs \$5.00 and they have, all told, only \$4.45. They need another \$5.00 - \$4.45 = \$.55 cents in order to be able to buy it.

All problems in arithmetic involve a similar form of reasoning which, as can be seen, is based on counting, grouping, and comparing. The art of arithmetic is designed to generalize and systematize this reasoning process—a process which converts real-world experiences into models (addition, subtraction, etc.). As Munzert (1991: 37) aptly puts it, arithmetical symbols and operations are designed to mirror our "contact with people and objects in the environment" (see also Gowan, Khatena and Torrance 1981, Lesgold 1988 and Norman 1993 on this point).

There are six basic arithmetical operations. These are: *addition, subtraction, multiplication, division, exponents,* and *roots*. Since arithmetic is a code, it comes as no surprise to find that these display patterns of paradigmatic and syntagmatic structure. The various patterns are called "laws" in arithmetic. For example, one law states that the product of multiplying any number by 0 is always 0:

$2 \times 0 = 0$

$3 \times 0 = 0$

$4 \times 0 = 0$

...

$n \times 0 = 0$

This is, of course, a paradigmatic (differentiating) feature of a particular sign in the code, since it entails assigning to 0 a differential meaning vis-à-vis all the other digits in the domain of multiplication. A second paradigmatic law states that division by 0 is not permitted or, more precisely, not defined. The expression $5 \div 0$, for instance, has no meaning. A third law asserts that the multiplication or division of two numbers with different signs (one positive and one negative) yields a negative product:

$2 \times -3 = -6$

$-4 \times 5 = -20$

$6 \times -7 = -42$

...

$n \times -m = -nm$

$-n \times m = -nm$

(where n and m are any digits, except 0)

$4 \div -2 = -2$

$-12 \div 3 = -4$

$-20 \div 4 = -5$

...

$n \div -m = -(n/m)$

$-n \div m = -(n/m)$

(where n and m are any digits, except 0)

A fourth syntagmatic law states that the multiplication or division of two negative numbers yields a positive number:

$-2 \times -3 = 6$

$-4 \times -5 = 20$

$-6 \times -7 = 42$

...

$-n \times -m = nm$

(where n and m are any digits, except 0)

$-4 \div -2 = 2$

$-12 \div -3 = 4$

$-20 \div -4 = 5$

...

$-n \div -m = n/m$

(where n and m are any digits, except 0)

A fifth syntagmatic law affirms that two numbers can be added in either order, and the sum will be the same. This is known as the *Commutative Law of Addition*:

$2 + 3 = 3 + 2 = 5$

$4 + 5 = 5 + 4 = 9$

$6 + 7 = 7 + 6 = 13$

. . .

$n + m = m + n$

(where n and m are any digits)

A sixth law states that numbers can be added by pairing them in any order, and the sum will be the same. This is known as the *Associative Law of Addition*. It can be illustrated with the addition of three digits:

$2 + 3 + 4 = (2 + 3) + 4 = 5 + 4 = 9$

Or

$2 + 3 + 4 = 2 + (3 + 4) = 2 + 7 = 9$

Generally

$n + m + o = (n + m) + o = n + (m + o)$

(where n, m, and o are any digits)

A seventh syntagmatic law asserts that numbers can be multiplied in either order, and the product will be the same. This is known as the *Commutative Law of Multiplication*:

$2 \times 3 = 3 \times 2 = 6$

$4 \times 5 = 5 \times 9 = 45$

$5 \times 7 = 7 \times 5 = 35$

. . .

$n \times m = m \times n$

(where n and m are any digits)

An eighth syntagmatic law states that numbers can be multiplied in any combination, and the final product will be the same. This is known as the *Associative Law of Multiplication*. This can be illustrated with three digits:

$2 \times 3 \times 4 = (2 \times 3) \times 4 = 6 \times 4 = 24$

Or

$2 \times 3 \times 4 = 2 \times (3 \times 4) = 2 \times 12 = 24$

Generally

$n \times m \times o = (n \times m) \times o = n \times (m \times o)$

(where n, m, and o are any digits)

Clearly, learning how to model arithmetical concepts in problems involves understanding these laws, which are based on common sense. For example, if we add the marbles on the left hand to those we have in our right hand the result will be same as adding them in reverse order. Moreover, since the operations are code-based, they form oppositions (as discussed in chapter 1). In this case, the two opposing poles of the opposition are reversals of each other: addition-versus-subtraction, multiplication-versus-division, and so on. The theory of opposition, as mentioned several times, is central to semiotic pedagogy, given its psychological validity (see, for example, Morris 1938, 1946, Osgood, Suci, and Tannenbaum 1957, Lévi-Strauss 1958, Greimas 1987, Sebeok and Danesi 2000). Signs of all kinds are recognizable as meaning-bearing structures in part through a perceivable difference built into some aspect of their physical constitution—a minimal difference in sound, a minimal difference in tone, a minimal difference in orientation, etc. The psychological importance of this fact was noticed first by the psychologists Wilhelm Wundt (1832–1920) and Edward B. Titchener (1867–1927), who were the ones to term it *opposition*. In the pedagogical domain, the theory of opposition suggests that we learn best when we envision things in oppositional ways: even-versus-odd, prime-versus-composite, addition-versus-subtraction, multiplication-versus-division, and so on, relating the two elements in the opposition structurally to each other. Signs acquire their meanings or values not in isolation, but in relation to other signs and to the codes in which they occur.

Teaching the Basic Operations

In order to teach arithmetical PS, we must start by showing how arithmetic is a modeling code. Consider the following simple arithmetical problem:

> [PROBLEM 2–2]
> A person has 8 pennies in the left hand and 4 in the right one. How many does the person have altogether?

The person has, of course, 12 pennies altogether. This problem shows that addition is, in effect, counting "consecutively." After counting 8 pennies in the left hand, we counted 5 more, reaching 12 as the end result. Of course, we could have done the reverse: that is, we could have started by counting the 4 pennies in the right hand and then 8 more. The result would have been the same. The number line is a perfect *model* of this counting procedure. A model in semiotics is any representamen that displays something in a concrete and easily understandable way. The line allows the teacher to show visually that the solution involves counting to the right of 0 in two steps: from 0 to 8 for "8 pennies," and then from 8 to 12 for the other "4 pennies:"

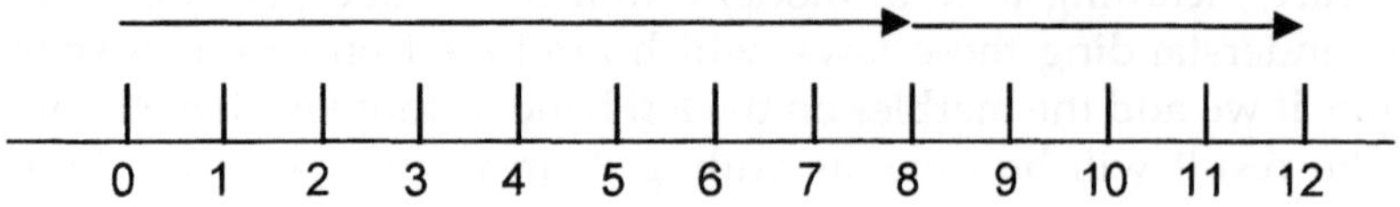

This is clearly a concretization strategy (discussed in the previous chapter). Note, as well, that the arrows are indexical devices that literally show the student the direction in which the counting process occurs. Addition notation is really shorthand for this kind of reasoning and modeling. At this point, the student can be shown how to convert the above diagram into appropriate addition notation and layout, such as: 8 + 4 = 12 or 4 + 8 = 12. What these allow us to do is avoid repeating the kind of reasoning just described over and over each time we add. In other words, as representamina they make addition mechanical and routine. The next step in the teaching process is naming the parts of the arithmetical layout. *Naming* is a mnemonic annotative strategy. It tells us what is going on. The numbers to be added together are called *addends*, no matter how they are laid out, and the result is called the *sum*:

8	+	4	=	12
↑	↑	↑	↑	↑
addend	*plus*	*addend*	*equals*	*sum*

Subtraction is the opposite of addition—the differential factor of the opposition being the direction of the counting process on the number line. For example, if we have 9 pennies in the left hand and we want to take away 5 from that hand and transfer them to the right hand, how many will we have left in the left hand? We will, of course, have 4 pennies left. "Taking away" is the basic concept underlying subtraction. It is indicated by the minus sign (-).To subtract 5 from 9, we count 5 "down" from 9, reaching 4 on the number line:

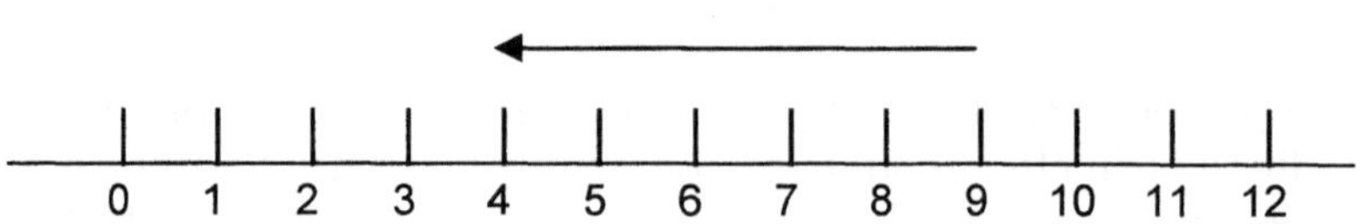

Subtraction notation and layout are shorthand for this kind of reasoning:

9	-	5	=	4
↑	↑	↑	↑	↑
minuend	*Minus*	*subtrahend*	*equals*	*difference*

The remaining two basic operations—multiplication and subtraction—are, really, condensed addition and subtraction operations, indicating two things: (1) that addition and subtraction are the only two basic operations, forming a perfect opposition; (2) that the other two operations are the result of using an economical method to stand for repeated addition and subtraction. Multiplication can thus be presented as follows:

2×3 means two 3's added together: $3 + 3$
4×3 means four 3's added together: $3 + 3 + 3 + 3$
5×3 means five 3's added together: $3 + 3 + 3 + 3 + 3$
...
$n \times 3$ means an indefinite number of 3's added together

At this point naming the parts is pedagogically effectual. The numbers multiplied together are called *factors* (as discussed in chapter 1). The result is called the *product*. The factor to be multiplied is also called the *multiplicand* and the factor that does the multiplying the *multiplier*. The product is also termed a *multiple* (of its factors), because it can be divided by any of its factors. In the multiplication $6 \times 2 = 12$, the product 12 can be divided by either 6 or 2. Thus, we say that 12 is a multiple of 6 and 2. Now that the conceptual structure of multiplication has been revealed, we can model the operation using the number line. For instance, $2 \times 3 = 6$ can be shown as 2 added to itself three times to the right of 0 on that line, reaching the end point 6:

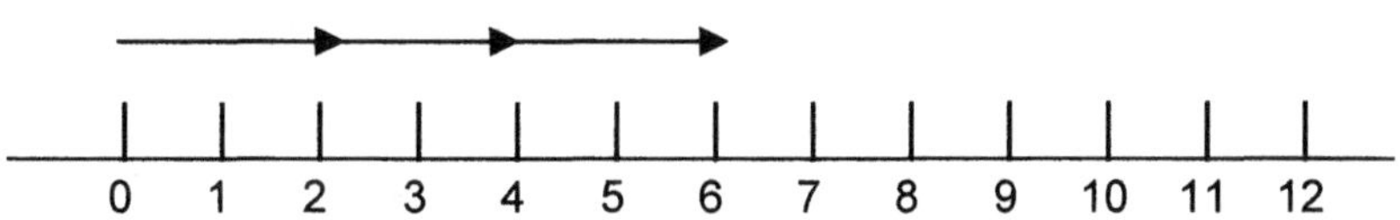

Notice that the same result would be obtained by adding 3 to itself twice, in conformity with the Commutative Law of Multiplication: $2 \times 3 = 3 \times 2$. Multiplication notation is shorthand for this kind of reasoning:

2	×	3	=	6
↑	↑	↑	↑	↑
factor	*times*	*factor*	*equals*	*product*

Now, division is repeated subtraction and the reverse of multiplication. The relevant research shows that this is indeed how students intuitively perceive it. A study of fourth- and fifth-graders by Graeber and Tirosh (1988) showed, for instance, that most defined division as "undoing multiplication." Children thus do not, initially, find division to be any harder conceptually than any other opposition (Mulligan and Mitchelmore 1997). Division can thus be presented as follows:

$6 \div 2$ means "How many times can 2 be taken away from 6?"
$8 \div 2$ means "How many times can 2 be taken away from 8?"
$10 \div 2$ means "How many times can 2 be taken away from 10?"
...
$n \div 2$ means "How many times can 2 be taken away from any number?"

The number line can be used again to model the repeated subtraction process. For example, $6 \div 3 = 2$ indicates that we must subtract 3 twice from 6:

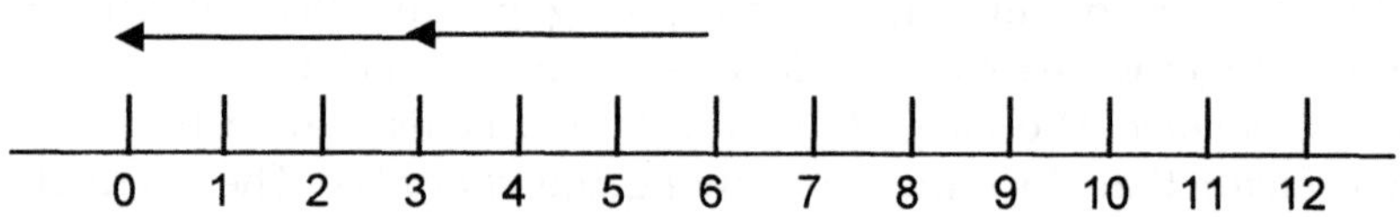

Since no "remainder" occurs from the division, it is said that 2 "divides evenly" into 6. When a number does not divide evenly into another, it leaves a remainder. For example, if we attempt to divide 153 by 7, using the number line method above, we will have to take away 21 and an additional length equivalent to 6/7 from 153. The numerator 6 is the remainder.

Division notation is shorthand for this kind of reasoning:

6	÷	2	=	3
↑	↑	↑	↑	↑
dividend	*divided by*	*divisor*	*equals*	*quotient*

From the foregoing discussion, it is obvious that the basic opposition of the arithmetical code is the addition-versus-subtraction one and that the others are derivatives of this opposition. As discussed in chapter 1, this shows that the arithmetical code has a basic binary opposition around which sets of other oppositions are derived or connected. The basic opposition, addition-versus-subtraction, can be called a first-order opposition, and the derived opposition, multiplication-versus-division, can be called a second-order opposition. The four oppositions thus form a square arrangement. In imitation of Greimas's 1987 semiotic square technique, this two-order four-part opposition can be shown as follows:

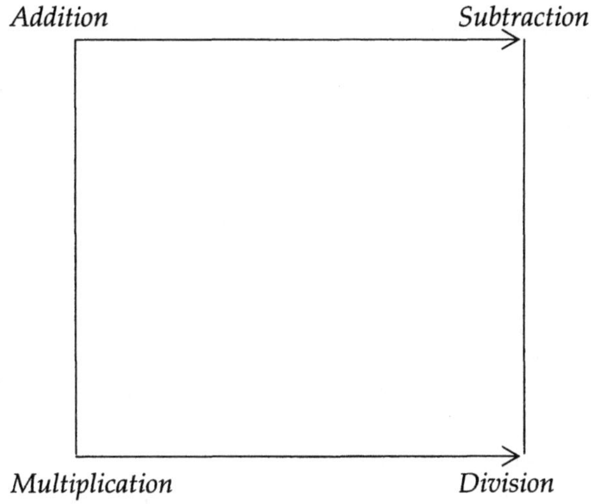

The left vertices (addition and multiplication) are the unmarked poles, whereas the right vertices (subtraction and division) are the marked poles—hence the reason for the arrowheads. This simple model confirms the psychological and structural validity of the traditional teaching order: addition-followed by-subtraction-followed by-multiplication-followed by division.

As an historical contextualization technique, it can be pointed out at this point that the number line model was the basis for a rather important discovery—the use of two number lines, crossing at right angles, for representing points in the plane, known as the Cartesian plane. The mathematician who discovered this simple opposition pattern, and after whom it is named, was René Descartes (1596–1650):

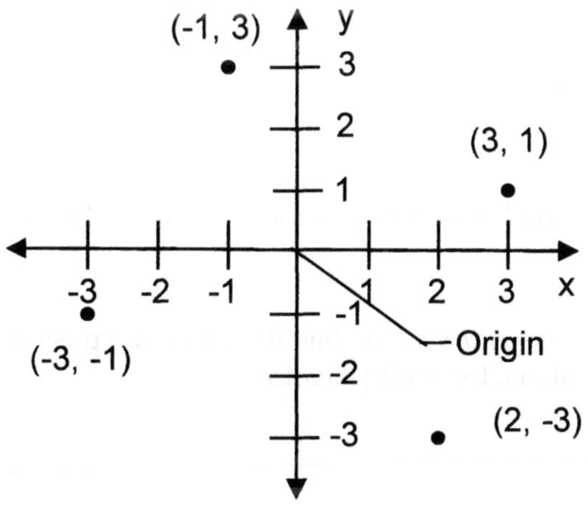

The point (3, 1), for instance, is defined by its location with respect to the two number lines, called *axes*: that is, it is located three equally-calibrated units to the right of the y-axis and one unit up from the x-axis; the point (2, -3) is located two units to the right of the y-axis and three units down from the x-axis; the point (-1, 3) is located one unit to the left of the y-axis and three units up from the x-axis; and the point (-3, -1) is located three units to the left of the y-axis and one unit down from the x-axis. In this modeling system, points are defined in relation to each other in terms of coordinates (which are the numerical points on each number line). Without going into the amazing subsequent discoveries that this simple modeling system has made possible since its introduction, not to mention the opening up of new branches of mathematics, suffice it to say that it reveals how a simple semiotic process—modeling something structurally (in oppositional terms)—leads to new vistas of knowledge and understanding.

Problems can now be easily designed to illustrate the notions presented in the above manner. Here are a few illustrative ones:

[PROBLEM 2–3]

Represent the following operations on a number line: 4 + 5, 2 +11

Semiotic Commentary: This type of problem is intended to get the student to understand the relation between number line models and addition layouts.

[PROBLEM 2–4]

Represent the following operations on a number line: 5 - 4, 8 - 3

Semiotic Commentary: This has the same purpose as [2–3] for subtraction.

[PROBLEM 2–5]

Represent the following operations on a number line: 2 × 4, 3 × 3

Semiotic Commentary: This has the same purpose as the previous problems for multiplication.

[PROBLEM 2–6]

Represent the following operations on a number line: 4 ÷ 2, 9 ÷ 3

Semiotic Commentary: This has the same purpose as the previous problems for division.

The same techniques discussed for teaching the arithmetic of positive integers apply to the teaching of the arithmetic of signed integers—that is, of positive and negative integers. As discussed in the previous chapter, the plus and minus signs are really signifiers of directionality, with the former representing right-hand movement and the latter left-hand movement on the number line. A positive number +n differs from its oppositional counterpart -n by being at the same point in distance from 0 but "on the other side." It forms a perfect indexical binary opposition. Adding and subtracting signed numbers can thus be modeled simply as directions to go left or right on the number line, according to the given sign. For example, adding (+2) and (+3), or simply 2 + 3, means go right two whole units from zero, and then three more to the right. The end point (or answer) is (+5):

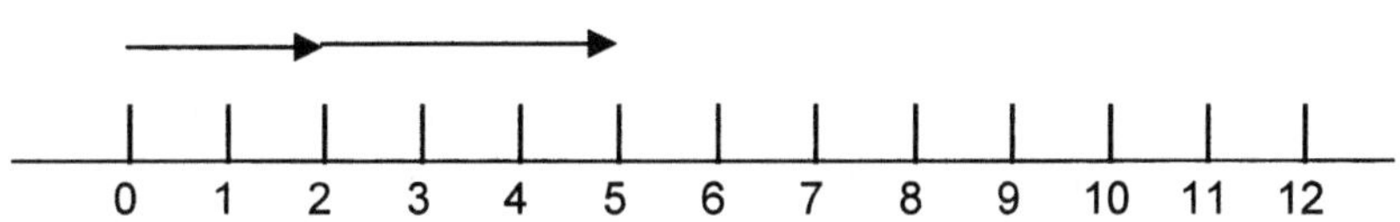

Adding two negative numbers can be modeled in an identical way. For example, (-3) + (-2) tells us to go left three units from 0 and then left two more. The end point (or answer) is (-5):

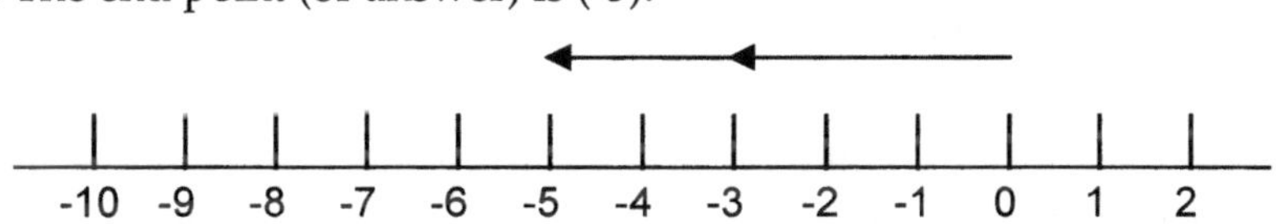

We point out to students that the sum (end point) of two negative numbers is also negative. To add numbers with different signs, the same type of reasoning applies. Consider (+5) + (-7). The first number tells us to go right five units from 0. The second tells us to change direction and go left seven units from there. The end point (or answer) is (-2):

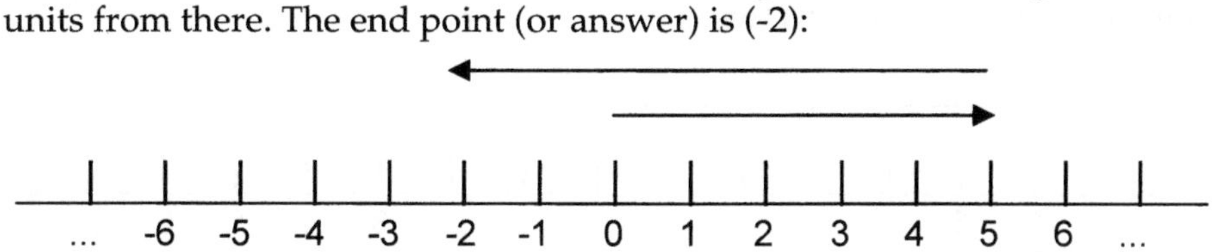

We can now point out to the student that the sign of the number with the larger absolute value (-7) is also the sign of the answer (end point). And the reason for this is evident on the number line. We can also indicate that adding a positive and a negative number is really a subtraction problem. If the signs are the same, we add the numbers and give the sum the common sign (as shown above). If the signs are different, we subtract the number with the smaller absolute value from the number with larger absolute value, assigning to the result the sign of the larger one. If a negative sign precedes a negative number, we change its sign to positive. We remind the learner as well that two negatives make a positive.

We can now go on to multiplication: for example, (+2) × (+3). Given the definition of multiplication above, this tells us to go right two units from 0 three times in sequence. The end result is the point (+6):

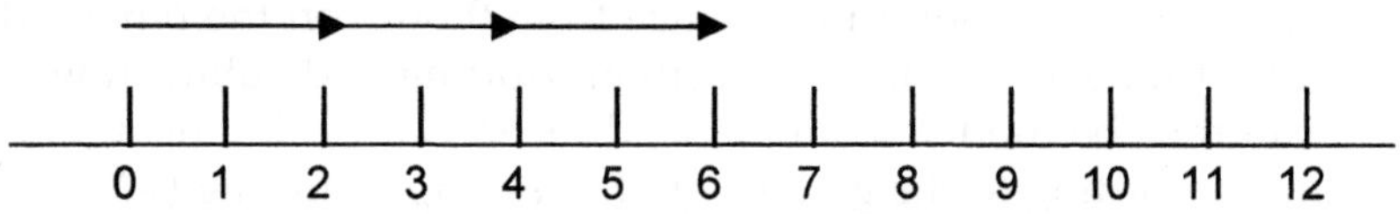

Next, consider the multiplication of (-2) × (-3), in which both numbers are negative. The best way to present this is in terms of the following steps:

- We can tell students to focus on the absolute values of both factors, namely |2| and |3|. This avoids potential confusion with the meaning of the negative signs.
- The end point of multiplying these two is, of course, |6|.
- On which side of 0 is |6| located?
- We remind students that two negatives make a positive. So it is located to the right at (+6).

What's the upshot of all this? If the signs of two factors in a multiplication are negative, their product is positive. The way this outcome can be modeled on the number line is slightly tricky. So, it is perhaps best to avoid this strategy since the above annotations should suffice.

Now, if the signs of the two digits are different, then their product is negative: for example, (-3) (+2) = (-6). This tells us that (-3) must go in the "same direction" (to the left) twice:

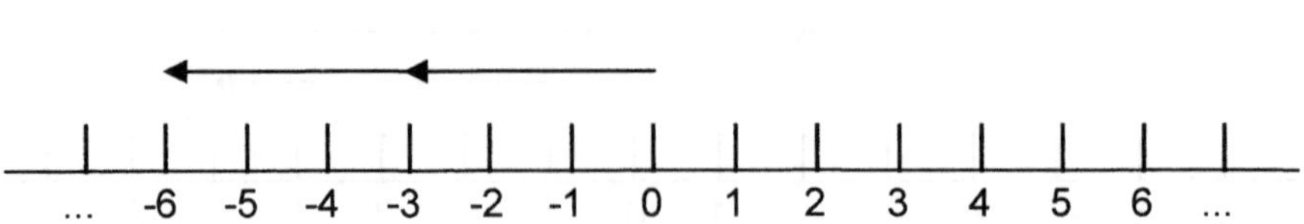

The basic reasoning used for the division of unsigned numbers applies to the division of signed numbers. We start by reminding the students that dividing two numbers with the same sign produces a positive quotient: (+12) ÷ (+4) = (+3), (-8) ÷ (-2) = (+4). Consider the former. It tells us that we have to subtract 4 units from (+12) three times:

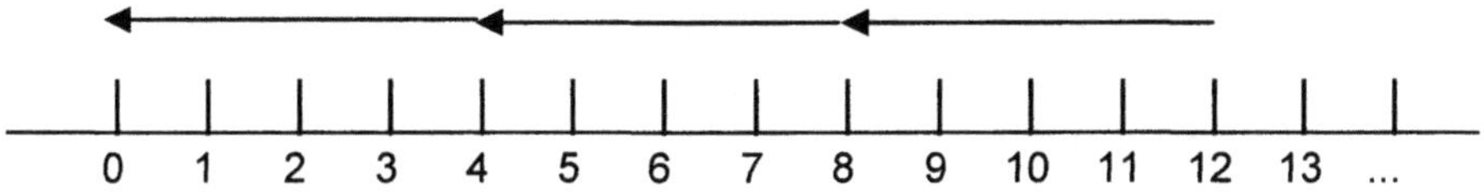

Dividing two numbers with different signs produces a negative quotient: (+12) ÷ (-4) = (-3). This means that the same movement on the number line is required, but that the negative sign reverses the direction. So, it is again more appropriate to use the step-by-step explanation technique discussed above (involving absolute values), rather than a number line modeling strategy.

At this point exemplification is required. Here are a few problems that can be easily set up to illustrate the above principles:

> [PROBLEM 2–7]
> Represent the following operations on a number line: (+4) + (+8), (+2) + (+18)
>
> *Semiotic Commentary:* This type of problem is intended to get the student to understand the relation between number line models and the addition of two positive numbers.
>
> [PROBLEM 2–8]
> Represent the following operations on a number line: (-5) + (-4), (-8) + (-3)
>
> *Semiotic Commentary:* This has the same purpose as [2–7] for two negative numbers.
>
> [PROBLEM 2–9]
> Represent the following operations on a number line: (+2) × (+4), (+3) × (+3), (+4) ÷ (+2), (+9) ÷ (+3)
>
> *Semiotic Commentary:* This has the same purpose as the previous problems for the multiplication and division of two positive numbers.

[PROBLEM 2–10]

Represent the following operations on a number line: $(-2) \times (-4)$, $(-3) \times (-3)$, $(-4) \div (-2)$, $(-9) \div (-3)$

Semiotic Commentary: This has the same purpose as the previous problems for the multiplication and division of two negative numbers.

[PROBLEM 2–11]

Represent the following operations on a number line: $(-2) \times (+4)$, $(+3) \times (-3)$, $(-4) \div (+2)$, $(+9) \div (-3)$

Semiotic Commentary: This has the same purpose as the previous problems for the multiplication and division of two oppositely-signed numbers. However, since division was not presented above with a number line model, the students will have to figure out how to model the examples by themselves. So, [2–11] has also an exploratory learning objective.

When the signed numbers and the basic operations have been taught, we can then teach students how to simplify complex operations. The signs in this case indicate the order in which the simplification is to be carried out. Parentheses ( ), brackets [ ], and braces { } are used to enclose numbers (and all kinds of mathematical symbols). They are called *signs of aggregation*. The enclosed symbols are called *expressions*. The order in which expressions are enclosed is: { [( )] }. To simplify an expression, we start with the innermost expression (the one in parentheses), then move to the next one (in square brackets), and finally to the expression enclosed by braces. Here is an example:

$$\{12 + [4 + 5 - (5 - 3) + 4] - 4\}$$

First, we simplify the innermost expression: $(5 - 3) = 2$, and plug this value into the larger expression, using the concretization technique of a box schema to highlight the expression in question:

$$\{12 + [4 + 5 - \boxed{(5 - 3)} + 4] - 4\} = \{12 + [4 + 5 - 2 + 4] - 4\}$$

Next, we simplify the bracketed expression [4 + 5 - 2 + 4] = 11 and plug this value in:

$$\{12 + \boxed{[4 + 5 - 2 + 4]} - 4\} = \{12 + 11 - 4\}$$

Finally, we simplify the resulting aggregate between braces. This produces the answer:

$$\{12 + 11 - 4\} = 19$$

We remind the student that a negative sign before an aggregate indicates that we must change the sign of the values within when removing the parentheses, brackets, or braces. Here are a few examples:

(1) $6 - (+8) = 6 - 8 = -2$
(2) $7 - (-8) = 7 + 8 = 15$
(3) $26 - (-13 + 12) = 26 - (-1) = 26 + 1 = 27$
(4) $-64 - [3 - (-1)] = -64 - [3 + 1)] = -64 - [+4] = -64 - 4 = -68$

Finally, the last thing to consider in teaching the elementary arithmetic of integers is factoring. We remind students that composite numbers are made up of factors. Factoring is, in effect, the operation of identifying the factors of such numbers. If a factor is part of two or more numbers, it is called (logically enough) a *common factor* of those numbers. For example, 1, 3, 5, and 15 are the factors of 15; and 1, 2, 4, 5, 10, and 20 are the factors of 20:

$$15 = 1 \times 15$$
$$15 = \boxed{1} \times 3 \times \boxed{5}$$

$$20 = 1 \times 20$$
$$20 = 1 \times 2 \times 10$$
$$20 = 1 \times 4 \times 5$$
$$20 = \boxed{1} \times 2 \times 2 \times \boxed{5}$$

As can be seen, 1 and 5 are *common* to both sets of factors. If two numbers have more than one common factor, the greatest one is called, logically, the *greatest common factor* (in the example above it is 5). It is also called the *greatest common divisor* since a factor of a number is also a divisor of that number. For example, the numbers 18, 30, and 48 have various common factors, as can be shown by decomposing the numbers into their factors. As can be seen, the greatest common factor (or divisor) in the three sets is 6:

$18 = 1 \times 18$
$18 = 1 \times 2 \times 9$
$18 = 1 \times 3 \times \boxed{6}$
$18 = 1 \times 3 \times 2 \times 3$

$30 = 1 \times 30$
$30 = 1 \times 2 \times 15$
$30 = 1 \times 5 \times \boxed{6}$
$30 = 1 \times 2 \times 3 \times 5$

$48 = 1 \times 48$
$48 = 1 \times 2 \times 24$
$48 = 1 \times 2 \times 2 \times 12$
$48 = 1 \times 2 \times 2 \times 2 \times \boxed{6}$

Problems exemplifying the foregoing operations can now be easily devised. The following are cases-in-point. They are self-explanatory and, thus, require no commentary:

[PROBLEM 2–12]
Evaluate the following:
(1) $7 - (-9 + 8)$
(2) $-7 [18 - (4 + 8)]$
(3) $-1 \{12 [-12 + (6 - 8)]\}$
(4) $(18 + 4) \{-3 [(4) (5 - 4)]\}$

[PROBLEM 2–13]
Break down the following two numbers into all their factors, indicating which is the lowest common factor:
(1) 16
(2) 90

[PROBLEM 2–14]
Find the greatest common factor (if any) of the following pairs of numbers:

(1) 27, 90
(2) 16, 80

Teaching Fractions and Decimals

Teaching the arithmetic of fractions involves using the same kinds of pedagogical principles and techniques discussed above and in the previous chapter. So, it is not necessary to go into any depth here. The first important notion is that of equivalence—since it is required for all the operations. The teacher can thus start by reminding students that fractions have the same value if the numerator and the denominator are multiplied by the same factor. Thus:

$$\frac{7}{5} = \frac{(7)(4)}{(5)(4)} = \frac{28}{20}$$

$$\frac{7}{5} = \frac{(7)(5)}{(5)(5)} = \frac{35}{25}$$

The fractions 7/5, 28/20, and 35/25 are called equivalent. The first one (7/5) is the default or unmarked fraction, because it is in its lowest terms. This means that the numerator and denominator are prime numbers. As always in the case of fractions, it is pedagogically propitious to use real-world models to exemplify the relevant concept. For example, it can be pointed out that 1/2 of a pie and 2/4 of the same pie are equivalent fractions, because they stand, as can be seen, for the same amount of pie:

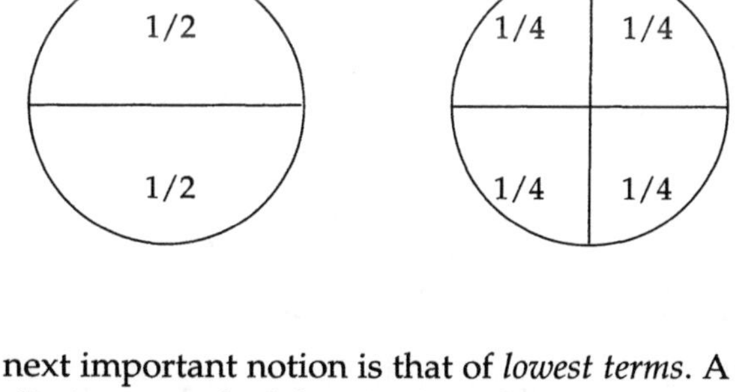

The next important notion is that of *lowest terms*. A fraction is in its lowest terms (in its unmarked form) when the numerator and the denominator are prime numbers and, thus, have no common divisor. So, to reduce 28/20 to its lowest terms, we do the opposite of what we did above. First, we factor the numerator and denominator:

$$\frac{28}{20} = \frac{(7)(4)}{(5)(4)}$$

Then, we eliminate the common factors, which can be shown by simply "crossing them out." This leaves the fraction in its lowest terms, because no more factoring is possible:

$$\frac{28}{20} = \frac{(7)\cancel{(4)}}{(5)\cancel{(4)}} = \frac{7}{5}$$

In actual fact, the process of elimination is really a short cut for division. The actual result of dividing the 4 in the numerator by the 4 in the denominator is 1, although this is not shown by convention.

Now, we come to the teaching of addition. To add two fractions with the same denominators, we simply add their numerators:

$$\frac{1}{4} + \frac{1}{4} = \frac{1+1}{4} = \frac{2}{4}$$

We can now reduce this to its lowest terms: 2/4 = 1/2. The reason why adding fractions with the same denominators is carried out in this way is an easy one to model with pie diagrams—one showing that it was cut in 1/2 and another in four equal parts. Adding two of those four parts, which is expressed arithmetically as 1/4 + 1/4, it is easy to see that the sum equals half of the pie:

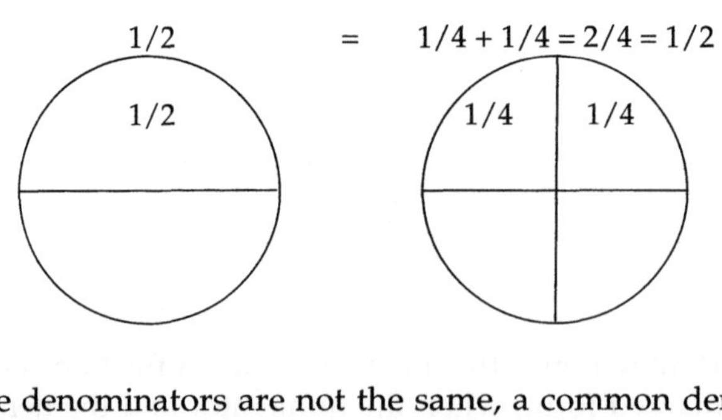

If the denominators are not the same, a common denominator must be found in order to add (or subtract) fractions. This can be done by multiplying the fraction with an integer that makes the denominators the same. Consider the following example:

$$\frac{5}{12} + \frac{2}{3} = ?$$

The second fraction, 2/3, is in its lowest terms. If we multiply its numerator and denominator by 4 we will get an equivalent fraction, 8/12:

$$\frac{2}{3} = \frac{(2)\,(4)}{(3)\,(4)} = \frac{8}{12}$$

We have now transformed the fraction into one with the same denominator of the first fraction, 5/12. Actually, 12 is the "least" possible value of the common denominator in this case. It is called, logically, the *least common denominator* (LCD). We can now add the two fractions:

$$\frac{5}{12} + \frac{8}{12} = \frac{5+8}{12} = \frac{13}{12} = 1\,1/12$$

The same reasoning applies to the addition or subtraction of fractions in their lowest terms. Consider the following subtraction problem:

$$\frac{3}{5} - \frac{2}{13} = ?$$

We can create a common denominator by multiplying the two denominators: 5 × 13 = 65. Thus, we multiply the denominator of the first fraction and its numerator by 13:

$$\frac{(3)\,(13)}{(5)\,(13)} = \frac{39}{65}$$

We multiply the denominator and numerator of the second fraction by 5:

$$\frac{(2)\,(5)}{(13)\,(5)} = \frac{10}{65}$$

We can now subtract the two fractions:

$$\frac{39}{65} - \frac{10}{65} = \frac{39 - 10}{65} = \frac{29}{65}$$

To multiply fractions, we simply inform the students to multiply their numerators and their denominators:

$$\frac{10}{7} \times \frac{1}{3} = \frac{(10)\ (1)}{(7)\ (3)} = \frac{10}{21}$$

Research has shown that multiplication is often confused with addition in the area of fractions. However, my own anecdotal classroom experiences suggest that if the four-part opposition order discussed above is followed in this case as well (addition-subtraction-multiplication-division) then confusion is attenuated.

We can now go on to signed fractions, which require no new conceptual learning. So, to multiply signed fractions, we instruct students to follow the usual rules for signed numbers. Here's an example:

$$\frac{10}{7} \times \frac{-1}{3} = \frac{(10)\ (-1)}{(7)\ (3)} = \frac{-10}{21}$$

Division is inverse multiplication (and the last pole in the four-part opposition). This means, literally, inverting the numerator and the denominator of one of the fractions and then multiplying the two:

$$\frac{8}{7} \div \frac{-3}{5} = \frac{(8)\ (5)}{(7)\ (-3)} = \frac{-40}{21}$$

After having taught the fractions, we can finish the instructional process by moving on to the arithmetic of decimals. Again, since there is nothing new conceptually to impart, students can be simply informed about what to do. To add and subtract decimals, all we must remind them to do is line up the decimal points and then proceed to add and subtract as usual, carrying over or borrowing as the case may be:

$$\begin{array}{r} 126.5 \\ \underline{+68.231} \\ 194.731 \end{array}$$

$$\begin{array}{r} 126.5 \\ \underline{-68.231} \\ 58.269 \end{array}$$

To multiply decimals, it is not necessary to align the decimal points. To determine the position of the decimal in the product we simply add the number of digits to the right of the decimal points in the factors. This is the number of decimal places required in the product. To divide decimal numbers, such as 62.744 ÷ 1.24, we instruct students to move the decimal point in the divisor to the right until the divisor becomes an integer:

1.24 → *Move the decimal over two places* → 124

Then, we tell them to move the decimal point in the dividend the same number of places:

62.744 → *Move the decimal over two places* → 6274.4

What we are really doing is multiplying the divisor (= the denominator) and the dividend (= the numerator) by the same value—one hundred in this case. This is tantamount to moving the decimal "two places to the right." Now, the operation of division can be carried out in the usual fashion.

Below are a few problems that can be designed to teach the operations discussed above. Their purposes are self-explanatory:

[PROBLEM 2–15]

Find the least common denominator of each set of fractions:

(1) 4/5, 9/10

(2) 1/2, 3/8, 9/16

(3) 5/6, 5/7, 4/9

(4) 1/3, 2/5, 5/7

[PROBLEM 2–16]

Carry out the following operations:

(1) 5/6 + 9/6

(2) 5/6 + 7/8

(3) (-7/6) - (-6/5)

(4) (6/7) × (-8/9)

(5) (2/3) ÷ (4/5)

[PROBLEM 2–17]

Evaluate the following:

(1) 4 [6.78 - (90.89 - 45.67)]

(2) 7.08 ÷ (-0.98)/4 [3 - (-4 - 5)]

(3) (67.98) {4.3 [4 + (-2) (-7.009)]}

(4) (-0.87) (-78.90) (-1.45)

Teaching Other Operations

Consider multiplying the number 3 fifteen times:

$$3 \times 3 \times 3 \times 3 \times 3 \times 3 \times 3 \times 3 \times 3 \times 3 \times 3 \times 3 \times 3 \times 3 \times 3 = 14{,}348{,}907$$

The computational task is clearly an unwieldy one. To make it more efficient mathematicians have come up with an abbreviation device called an *exponent*. An exponent is a small number or letter written as a superscript over a *base* (the number that is to be multiplied by itself). For example, with this device the above multiplication can be compressed into the form 3^{15}, in which the base is 3 and the exponent is 15. This notation tells us that 3 is to be multiplied by itself fifteen times. In general, n^m is an exponential number indicating that a number n is to be multiplied m times.

Exponential notation can be used with integers (positive and negative), fractions, decimals, and other types of numbers. Here are some examples:

$(-4)^3 = (-4) \times (-4) \times (-4) = -64$

$(1)^5 = (1) \times (1) \times (1) \times (1) \times (1) = 1$

$(.03)^4 = (.03) \times (.03) \times (.03) \times (.03) = 0.0000008$

The term *power* is often used. For example, 3^5 is read the "fifth power of three" or "three to the power of five." The second and third powers of a number are called its *square* and its *cube*:

3^0 = *Three to the power of zero* = 1

3^1 = *Three to the power of one* = 3

3^2 = *Three to the power of two* = *Three squared* = 9

3^3 = *Three to the power of three* = *Three cubed* = 27

3^4 = *Three to the fourth power* = 81

3^5 = *Three to the fifth power* = 243

...

3^n = *Three to the nth power*

This notation has led to several truly fascinating discoveries, as will be discussed in the next chapter: for instance, the zero power of a number is always equal to one ($n^0 = 1$), and the first power of a number is the number itself ($n^1 = n$). It has also led to a series of related concepts and techniques. One of these is finding the *root*, which is the inverse or opposite operation. It consists in discovering what the base of a certain power is. For example:

$4^2 = 16$

Or

$4 \times 4 = 16$

The root of 16 is thus 4. It is also called the "square root" of 16. Another square root of 16 is -4 because the product of two negative numbers is positive:

$-4 \times -4 = 16$

The root sign, called a *radical*, is √. The number of times the root is taken as a factor is called the *index*—indicated as a superscript number inserted within the radical. Basically, it tells us what kind of root is required (square, cube, etc.). The number enclosed by the radical is called the *radicand.* So, the square root of 16 is shown as follows:

$\sqrt[2]{16} = \pm 4$ (read: *The square root of 16 is positive or negative* 4)

By convention, when the index is 2 (as it is above), it is not written:

$\sqrt{16} = \pm 4$

Below are examples of other kinds of roots:

$\sqrt[3]{27} = 3$ (read: *The cube root of 27 is 3*)

$\sqrt[4]{625} = 5$ (read: *The fourth root of 625 is 5*)

$\sqrt[5]{32} = 2$ (read: *The fifth root of 32 is 2*)

What do these stand for, the student might ask? Let's take each one separately. Finding the cube root means finding a base number which, when multiplied by itself three times, produces 27. That root is, clearly, 3:

$3^3 = 3 \times 3 \times 3 = 27 \rightarrow$ *Reversing the operation* $\rightarrow \sqrt[3]{27} = 3$

Finding the fourth root means finding a number which, when multiplied by itself four times, produces 625. That number is 5:

$5^4 = 5 \times 5 \times 5 \times 5 = 625 \rightarrow$ *Reversing the operation* $\rightarrow \sqrt[4]{625} = 5$

And, finding the fifth root means finding a number which, when multiplied by itself five times, produces 32 as a result. That number is 2:

$2^5 = 2 \times 2 \times 2 \times 2 \times 2 = 32 \rightarrow$ *Reversing the operation* $\rightarrow \sqrt[5]{32} = 2$

At this point, problems designed to reinforce the notions just imparted are easy to construct. The following are examples:

[PROBLEM 2–18]
Use exponents to represent the following multiplications:
(1) $(2 \times 2 \times 2)$
(2) $(8 \times 8 \times 8 \times 8 \times 8 \times 8)$
(3) $(9 \times 9 \times 9 \times 9 \times 9 \times 9 \times 9 \times 9 \times 9 \times 9 \times 9 \times 9)$
(4) $(a \times a \times a \times a \times a \times a \times a \times a \times a)$
(5) $(s \times c)\ (s \times c)\ (s \times c)\ (s \times c)\ (s \times c)$

Semiotic Commentary: This type of problem is designed to get the student to represent repeated multiplication with an exponential number. This is a basic sign-making process. Example (5) adds a little openness to the problem-solving process, given its layout.

[PROBLEM 2–19]
Show the factors that the following represent:
(1) 5^5
(2) 12^6

(3) x^9
(4) x^y

Semiotic Commentary: This type of problem is the inverse of [2–18]. Example (4) adds a little openness to the problem-solving process, since it requires students to use representational know-how on their own.

[PROBLEM 2–20]
What is the value of each of the following?
(1) $\sqrt{9}$
(2) $\sqrt[3]{8}$
(3) $\sqrt[4]{16}$

Semiotic Commentary: This type of problem simply provides practice in root extraction.

The topic of exponents brings out the power of new signs in knowledge-making and discovery. Initially, exponential notation had a simple purpose (as we saw)—to represent the multiplication of identical digits with a new condensed notation. But, as mathematicians used the notation they started discovering new and unexpected things, such as, for example, that n^0 is equal to 1. Without going into the simple proof of this here, suffice it to say that this property of 0 would never have been discovered without the notation. Exponential notation has also led to the discovery of other laws of arithmetic (as we shall see in the next chapter). To the semiotician, this is really nothing unexpected. *Structural economy*, as it can be called (Danesi 2007b: 34–45), characterizes all kinds of human semiotic activities and leads to discoveries by suggesting new links between signifiers and their signifieds. It is a virtual law of language change. The English spoken and written today is structurally much more economical than was Old English, which had a full-fledged case system that has practically disappeared from the modern language. Analogously, the case system of Latin no longer exists in its modern day descendants (Italian, French, Spanish, etc.). In these languages, sentence length has been greatly reduced by such elimination—a phenomenon that occurred initially because the grammatically-significant endings of words were not pronounced or written, leading to their abbreviation or elimination.

Economy often correlates with frequency—the more frequently a form or category is used the more likely it is that it will be replaced by a reduced equivalent. This is why common abbreviations and acronyms are used for common ideas: for example, *ad* (advertisement), *photo* (photograph), *NATO* (North Atlantic Treaty Organization), *laser* (light amplification by stimulated emission of radiation), *DVD* (digital video disc), *it's* (it is), *UN* (United Nations), *CD* (compact disc), and so on. These save effort. Reduction is also a characteristic style in the creation of tables, technical and scientific material, indexes, footnotes, and bibliographies. The features evident today in text messages—*U2* (you too), *How R U* (How are you?), *b4* (before), *f2f* (face-to-face) *gr8* (great), *g2g* (gotta go), *ezy* (easy), *iluvu* (I love you), *ruok* (Are you OK?)—display an analogous tendency to reduce the length of frequently-used words, phrases, and sentences.

Psychologically, manifestations of structural economy suggest the presence of a Principle of Least Effort (PLE) in human semiotic behavior, which can be defined simply as the propensity to reduce the physical structure of signs and sign systems so as to minimize effort. The PLE came to the forefront in the 1930s and 1940s, when the Harvard linguist George Kingsley Zipf (1902–1950) published a series of key studies showing that the most frequently-used words in written texts (newspaper articles, novels, etc.)—*a, the, and,* etc.—were also the shortest in length (in terms of the number of alphabet characters that constituted them). The higher the frequency of a word, the more likely it was, Zipf concluded, to show an economy of form (fewer letters, fewer syllables, etc.). More technically, Zipf discovered that there was an inverse correlation between the length of a specific word (in number of phonemes) and its rank order in the language (its position in order of its frequency of occurrence in texts of all kinds). The higher the rank order of a word (the more frequent it was in actual usage), the more it tended to be reduced in form. Articles, conjunctions, and other such forms, which had a high rank order, were typically monosyllabic, consisting of 1 to 3 phonemes (see, for example, Zipf 1935, 1949).

To grasp the substance of "Zipf's Law," as Zipf's discovery came to be called, all one has to do is take the words in a substantial corpus of text, such as an issue of the *New York Times*, and count the number of times each word in it appears. Variants of the same basic word are to be counted as separate tokens of the word: for example, *play, played, playing* all count as three manifestations of the same word. If one were then to plot the frequencies of the words on a histogram and sort them by frequency rank, the curve will approach a straight line with a slope of -1. The exact same type of result emerges from counting words in textbooks, recipe collections, speeches, novels, and all other verbal texts. The larger the corpus the more the curve tends towards this slope. In effect, the more frequent a word in a text the shorter it tends to be. Using data from widely-divergent languages, Zipf found, moreover, that this was true of any language and, not only words, but also grammatical systems, Chinese characters, and other semiotic forms and systems. Some linguists have rejected the PLE, suggesting that Zipfian curves are simple consequences of his particular method of regarding a message source as a random process and, thus, a function of probability considerations, not psychological propensities. But many others have accepted the PLE as a general psychological framework for explaining such everyday phenomena as abbreviation and acronymy in language.

For the sake of historical accuracy, it should be mentioned that some of Zipf's ideas were prefigured by various scientists in the nineteenth century. For example, the American astronomer Simon Newcomb (1835–1909) found, in 1881, that if the digits used for a task are not entirely random, but somehow socially based, the distribution of the first digit is not uniform—1 tends to be the first digit in about 30% of cases, 2 will come up in about 18% of cases, 3 in 12%, 4 in 9%, 5 in 8%, etc. Newcomb came to this discovery by noticing that the first pages of books of logarithms were soiled much more than the remaining pages. In 1938, physicist Frank Benford (1883–1948) investigated listings of data, finding a similar pattern to the one uncovered by Newcomb in income tax and population figures, as well as in the distribution of street addresses of people listed in phone books. Zipf's main contribution was in showing empirically that patterns of this type manifest themselves regularly and "blindly" in all kinds of human representational efforts, especially in language. Shortly after the publication of Zipf's ideas, the mathematician Benoit Mandelbrot (b. 1924), who developed the modern-day branch of mathematics known as fractal geometry, became fascinated by them, detecting in them an indirect confirmation of what is called a "scaling" law in biology. Mandelbrot made significant mathematical modifications to Zipf's original Law—modifications that are used to this day to study structural economy phenomena in language and other sign systems.

The basic implication to be derived from Zipfian analysis can be reformulated in semiotic terms as the Signifier Compression Principle (SCP). This claims more specifically that the tendency in semiosis is towards the compression of signifiers, while the meaning range of signifieds is preserved and, in some cases, even expanded (Danesi 2006b). The SCP is, in some cases, related to time. The longer words and expressions and have been around, for instance, the more likely they will be for the simple reason that they are used frequently. However, this is not always the case. Often, simple convenience is the factor leading to compression. The abbreviation *PC* = *personal computer* was used almost the instant the device entered into usage. Simply put, compression is the outcome of the tendency to reduce effort, either because of usage over time or because of immediacy of utilization.

The foregoing discussion was meant to bring out the fact exponential notation and other symbolic constructs of math representation have generated new ways of understanding something. Perhaps in no other domain of human knowledge does Zipf's Law operate so frequently and has revealed so many facets of hidden structure than in mathematics.

Arithmetical Competence

To summarize the foregoing discussion, it can be said that an arithmetical problem (in addition, subtraction, etc.) is fundamentally a semiotic model representing counting notions of various kinds in some sign-based way. Arithmetical competence can, thus, be defined as the ability to represent counting notions (the signifieds) with appropriate arithmetical models (the signifiers). The representational process unfolds, as we have seen, at various levels: from a rudimentary one in which the arithmetical models represent simple counting procedures, to advanced ones in which operations are represented with condensed forms, such as exponential notation. The basic intent of semiotic pedagogy is to get students to convert their intuitive knowledge of counting into such representational forms. The techniques used are the usual ones: concretization (the use of visual and diagrammatic strategies), annotation (the use of appropriate commentary indicating the structural principles involved), contextualization (the use of real-world examples to illustrate the principles), generalization (the use of abstract representamina to summarize something in general), the design of problems allowing the student to manipulate the structural principles involved, and so on. The "flow" of teaching is dictated by considering the oppositions involved in arithmetic. The semiotic square discussed above for the teaching of the basic operations is a case-in-point. It suggests that we should always start with a primary opposition (and its unmarked pole) and move sequentially to higher order oppositions. This principle of teaching finds its validity not only in actual classroom experience, but also in psychological research findings on learning, even though psychologists typically do not identify it as such (Rosch 1973a, 1973b).

The arithmetical problems constructed in this chapter for pedagogical purposes have been (essentially) closed texts designed to get students to learn to recognize and use some structure or principle. However, it is in the use of more open problem texts that students will come to grasp arithmetical structure more effectively and permanently through exploration. One such open textual type is the so-called *alphametic* puzzle. This is an ingenious type of arithmetical problem in which each digit has been replaced by a specific letter. Solvers are required to reconstruct the original numerical layout on the basis of their knowledge of arithmetical structure (Danesi 1982). To the best of my knowledge, it was puzzlist J. A. H. Hunter (1965a: 119) who first used the word *alphametic* to distinguish it from a cryptarithmetical puzzle (such as the one discussed in the previous chapter), because the letters replacing the numbers in it form words, phrases or sentences, rather than just random sequences (Brooke 1969). The originator of alphametics was, however, not Hunter, but none other than the great English puzzlist, Henry E. Dudeney (1847–1930), who referred to them as exercises in "verbal arithmetic." Alphametics are quite popular, as witnessed by their appearance in newspapers and magazines. Pedagogically they are extremely useful because they presuppose that students have grasped fundamental principles of arithmetical structure and how it is represented.

Here is Dudeney's original alphametic:

[PROBLEM 2–21]

$$
\begin{array}{ccccc}
 & S & E & N & D \\
+ & M & O & R & E \\
\hline
M & O & N & E & Y
\end{array}
$$

It is worthwhile going through a detailed discussion of its solution, for this will bring out the essence of what arithmetical competence is all about. Since it is a problem in addition, it can instantly be established that the M at the extreme left is a carry-over digit equal to 1, because 1 is the only carry-over possible when two digits are added together in the previous column—in this case S + M—even if that column has itself a carry-over from the column before it. If the student does not see this, then it is necessary to get the student to consider the column in question in a concrete fashion, plugging in actual values for S, M, and O. There are 10 digits including zero. The maximum two digits can add up to is 17; this would occur when the two largest digits, 9 and 8, are added together. So, if we let the two digits in the column equal 9 and 8, just for the sake of concrete illustration, ignoring for the moment the actual letters that are there, we would get the following. We tell the student to note that there is no carry-over in this case because a carry-over would make the 0 = 8, and this is impossible since that number has already been assigned (to M):

$$
\begin{array}{ccccc}
 & 9 & E & N & D \\
+ & 8 & O & R & E \\
\hline
1 & 7 & N & E & Y
\end{array}
$$

Incidentally, this demonstrates that M = 1. So, we can put its numerical value in the puzzle, noticing that it occurs in two places:

$$
\begin{array}{ccccc}
 & S & E & N & D \\
+ & 1 & O & R & E \\
\hline
1 & O & N & E & Y
\end{array}
$$

Now, we consider again the S + 1 = O column because it has in it a decoded number. Since M = 1, we know that O does not equal 1. It must therefore equal 0, since the S above it can only be, at maximum, equal to 9. Adding together 9 + 1 we would get 10. We cannot get 11 because that would make the O below equal to 1, which, as we have just determined, is the value of M. So, let's replace O = 0 in the places where it is found in the layout:

	S	E	N	D
+	1	0	R	E
1	0	N	E	Y

Now, the S must be either 9 or 8. The 8 possibility would be required if there were a carry-over from the third-left column (E + 0 = N). But there can be no carry-over, even if we let E = 9, because that column has a 0 in it. Why? Because that would mean that N = 1: 9 (= E) + 0 + 1 (carry-over) = 1 (= N). However, the digit 1 has been already been assigned to M. So, there can be no carry-over from the E + 0 = N column to the S + 1 = 0 column. We thus conclude that S = 9 by the process of elimination:

	9	E	N	D
+	1	0	R	E
1	0	N	E	Y

Next, we turn our attention to the E + 0 = N column. It must have a carry-over from the previous column, because if there were no carry-over, this column would not make arithmetical sense. Any number added to 0 would produce that number as a sum: 2 + 0 = 2, 3 + 0 = 3, etc. So, without a carry-over, E + 0 should add up to E. But it does not. So, E + 0 = N must have a carry-over:

		1		
	9	E	N	D
+	1	0	R	E
1	0	N	E	Y

Let's write this out separately from the layout:

1 + E + 0 = N

1 + E = N ← *Simplifying*

N = E + 1 ← *Rearranging*

We leave this for the moment and go on to consider the N + R = E column. As we have just determined, the addition of N + R produces a carry-over. In arithmetical terms, this means that the sum of N + R, which is E, is greater than 10. This fact can be represented, of course, as E + 10. So, the column can be represented as:

(1) $N + R = E + 10$

This assumes, however, that there is no carry-over from the previous column. If there is a carry-over from that column, then the appropriate equation is:

1 (carry-over) $+ N + R = E + 10$
(2) $N + R = E + 9$ ← *Simplifying and Rearranging*

We have established that $N = E + 1$. So, we can now substitute this value in the two possibilities for the column just considered:

(1) *With no carry-over:*
$N + R = E + 10$
$E + 1 + R = E + 10$ ← *Replacing N with E + 1*
$R = 9$ ← *Simplifying*

(2) *With a carry-over:*
$N + R = E + 9$
$E + 1 + R = E + 9$ ← *Replacing N with E + 1*
$R = 8$ ← *Simplifying*

Therefore, R = 9 or 8. We have already established that S = 9. So, we can conclude that R = 8 by the process of elimination:

	9	E	N	D
+	1	0	8	E
1	0	N	E	Y

Now, let's consider the N + 8 = E column again. The digits 0 and 1 have already been assigned, so we can deduce that this column produces a carry-over, since N is at minimum 2 (not 0 or 1). Adding 2 to 8 would produce a carry-over (as would any other higher number). Now, it is obvious that the

D + E = Y column will also produce a carry-over, since N or E also cannot be 0 or 1 in the N + 8 = E column. It is to avoid this that a carry-over is needed.

N also cannot be 2, for then E would be equal to 0 or 1 in the N + 8 = E column, which is impossible since these two digits have already been assigned. Students can try this out for themselves if they do not see it. So, we should try out a few scenarios, starting with N = 3. We can represent hypothetical replacements in some different font or style (such as underline):

$$\begin{array}{ccccc} & 9 & E & \underline{3} & D \\ + & 1 & 0 & 8 & E \\ \hline 1 & 0 & \underline{3} & E & Y \end{array}$$

From the E + 0 = 3 column, we can see that E must equal 2, given that the column has a carry-over:

$$\begin{array}{ccccc} & & 1 & & \\ & 9 & \underline{2} & \underline{3} & D \\ + & 1 & 0 & 8 & \underline{2} \\ \hline 1 & 0 & \underline{3} & \underline{2} & Y \end{array}$$

So, the D in the right-most column must be 8 or 9, since only 8 (= D) + 2 or 9 (= D) + 2 will produce the carry-over. But this is not possible, since these digits have been already assigned. So, we discard the original hypothesis, namely that N = 3. We can now try N = 4 instead:

$$\begin{array}{ccccc} & 9 & E & \underline{4} & D \\ + & 1 & 0 & 8 & E \\ \hline 1 & 0 & \underline{4} & E & Y \end{array}$$

This replacement makes E = 3:

$$\begin{array}{ccccc} & & 1 & & \\ & 9 & \underline{3} & \underline{4} & D \\ + & 1 & 0 & 8 & \underline{3} \\ \hline 1 & 0 & \underline{4} & \underline{3} & Y \end{array}$$

Now, D must be equal to 7, 8, or 9 to produce the carry-over. We have already assigned 8 and 9; But if D = 7, then Y = 0, since 7 (= D) + 3 = 10. But 0 has already been assigned. Once again, we discard the working hypothesis that N = 4, and go on to N = 5 as our next hypothesis:

$$\begin{array}{ccccc} & 9 & E & \underline{5} & D \\ + & 1 & 0 & 8 & E \\ \hline 1 & 0 & \underline{5} & E & Y \end{array}$$

This replacement makes E = 4:

$$\begin{array}{ccccc} & & 1 & & \\ & 9 & \underline{4} & \underline{5} & D \\ + & 1 & 0 & 8 & \underline{4} \\ \hline 1 & 0 & \underline{5} & \underline{4} & Y \end{array}$$

This means that D must be 6, 7, 8, or 9. We can discard 8 and 9, since these digits have already been assigned. We can also discard D = 7, because this would make Y = 1, and this digit too has been assigned. And we can reject D = 6, because this would make Y = 0, which has also been assigned. So, we reject the hypothesis that N = 5, and go on to N = 6 as our next hypothesis:

$$\begin{array}{ccccc} & 9 & E & \underline{6} & D \\ + & 1 & 0 & 8 & E \\ \hline 1 & 0 & \underline{6} & E & Y \end{array}$$

This replacement makes E = 5:

$$\begin{array}{ccccc} & & 1 & & \\ & 9 & \underline{5} & \underline{6} & D \\ + & 1 & 0 & 8 & \underline{5} \\ \hline 1 & 0 & \underline{6} & \underline{5} & Y \end{array}$$

This means that D = 7, 8, or 9. As before, we discard 8 and 9, because these have already been assigned. However D = 7 works perfectly, since 7 (= D) + 5 = 12. This means, of course, that Y = 2. The solution is now complete:

$$\begin{array}{ccccc} & 9 & 5 & 6 & 7 \\ + & 1 & 0 & 8 & 5 \\ \hline 1 & 0 & 6 & 5 & 2 \end{array}$$

As this example shows, alphametics lay bare the structure of arithmetical problems, involving a play on this very structure. The process is fundamentally semiotic—letting one thing stand for another and deducing the substitutes through structural reasoning. Solving alphametics also entails a cross-systemic analysis, since it involves knowing the structural features of the orthographic and grammatical code in question. In French, for instance, most words end in a consonant, and therefore, as in the case of English alphametics, most numerical layouts will contain non-identical digits in the units position. If identical digits occur then this might suggest plural forms of nouns, adjectives and determiners (articles, demonstratives, etc.). Italian, on the other hand, is a vocalic language. Almost all of its words end in a vowel. For this reason, identical digits appear frequently in the units position.

Clearly, alphametics add an exploratory component to the learning of arithmetical PS. As an historical aside, it can be mentioned to students that the study of series came about as the result of a similar exploration with arithmetical structure. It is worth repeating it here because it shows again that representation and discovery are consistently intertwined. The story goes that the great Karl Friedrich Gauss was only ten years old when he purportedly dazzled his math teacher, after the teacher had asked the class to cast the sum of all the numbers from one to one hundred, laying them out in the usual vertical way:

$$\begin{array}{l} 1 \\ 2 \\ 3 \\ 4 \\ \ldots \\ \underline{+\,100} \\ ? \end{array}$$

Gauss raised his hand within seconds, giving the correct response of 5,050. When his teacher asked little Karl how he was able to come up with the answer so quickly, he is said to have replied (more or less) as follows:

> I laid out the numbers in order, removing from the layout the middle number 50 and the last number 100: {1, 2, 3, …, 97, 98, 99}. In this layout there are 49 pairs of numbers that add up to one hundred. The pairs are made up as follows: the first and last numbers in the layout (1 + 99 = 100), the second and second-last numbers (2 + 98 = 100), the third and third-last (3 + 97 = 100), and so on. That makes 4,900, of course. Adding to this the 50 and 100 that were removed gives 5,050.

In effect, by changing the vertical layout into a horizontal one, and pairing numbers systematically, Gauss had discovered and proved how to sum an arithmetical series: {1 + 2 + 3 +…+ 100}. The problem given to Gauss and his classmates can be expressed more generally as follows: What is the sum of n consecutive numbers: {1 + 2 + 3 + … + n}? The answer is: n(n + 1)/2. Plugging 100 into n produces the answer:

$$\frac{n(n+1)}{2} = \frac{100(100+1)}{2} = \frac{(100)(101)}{2} = \frac{10{,}100}{2} = 5{,}050$$

Let's put down what Gauss did in his head, with a few slight modifications. First, we lay out the terms in numerical order and then in reverse order, putting one under the other:

(1)	1	2	3	…	100
	↕	↕	↕		↕
(2)	100	99	98	…	1

Next, we add the two numbers in each column together:

(1)	1	2	3	…	100
+	+	+	+		+
(2)	100	99	98	…	1
Sum	101	101	101	…	101

This makes it obvious that the sum in each case is the same—101. How many times do we get this sum? Since there are one hundred columns, that sum occurs 100 times. Thus, the overall sum of the columns is: 100 × 101 = 10,100. Now, this is twice the sum of all the integers from 1 to 100, because we added two series together—the top is the initial series in numerical order and the bottom the same series in reverse order. So, all we have to do is divide the overall sum in half: 10,100/2 = 5050.

Now, let's generalize this process. We note that 101 is one more than the number of terms in the series, 100. Thus, if n is the number of terms in a series, (n + 1) stands for one more than n and the overall sum S of the series can be represented as:

$$\frac{n \times (n+1)}{2} = S$$

Since solving alphametics involves a play on structure, they engage the student in a similar Gaussian form of reasoning, as it can be called. They are thus extremely useful for building arithmetical competence, allowing students to grasp the fundamental relation between representation and arithmetical structure. They also interrelate mathematics and language. It can, thus, also be mentioned that mathematical and linguistic structure share many features. What mathematicians do when they factor out the term xy in the expression axy + bxy = cxy is similar to what linguists do when they identify the form *-berry* as a recurrent feature in words such as *blueberry*, *cranberry*, etc. Some linguists and mathematicians have, in fact, overtly adopted each other's techniques (Hockett 1967, Brainerd 1971, Devlin 1998, 2000). Actually, the connection between math, language, and the humanities generally has been felt to be an intrinsic one since antiquity. The three constituted a liberal arts education—a form of education considered to be essential for a free human being in ancient Greece and Rome, including grammar, logic, rhetoric, mathematics, and music. Colleges in the United States that concentrate on teaching these disciplines are still called liberal arts colleges.

Given the intrinsic connection between language and math, it is worthwhile considering one other type of arithmetical puzzle—a type that uses ordinary language to deceive the solver into perceiving the conditions given as purely arithmetical, whereas they are not. Puzzles of this type can be called simply *contrivances*. Their main purpose is to confuse the problem-solver and language is used, as Churchman (1970: 151) puts it, with a "malicious intent" to deceive.

There are two types of contrivances. The first employs a contrived arithmetical problem style to hide or obscure the meanings of lexical items (Jacoby 1962):

> [PROBLEM 2–22]
> How many cubic inches of dirt are there in a hole that is one foot deep, two feet wide and six feet long?

> [PROBLEM 2–23]
> How many times can you subtract the numeral one from the numeral twenty-five?

> [PROBLEM 2–24]
> A train leaves New York for Chicago traveling at the rate of 100 miles per hour. Another train leaves Chicago for New York an hour later traveling at the rate of 75 miles per hour. When the two trains meet, which one is nearer New York?

These puzzles intentionally contain irrelevant information (mainly in the form of numbers), which directs the problem-solver's attention away from word meanings. Thus in [2–22] the meaning of *hole* may not be directly obvious, because of the style in which the puzzle is cast—a *hole* cannot contain anything. In [2–23], the meaning of *numeral* permits only one subtraction of the numeral *one* from the numeral *twenty-five*. Problem [2–24] simulates the style of a rate problem. It thus obscures the meaning of the verb *meet*. Obviously, when two trains *meet*, neither one is nearer to New York.

The second type of contrivance has one general characteristic: it provides arithmetical information which "pulls" the problem-solver's attention away, so to speak, from interpreting it correctly (Jacoby 1962: 15, Mott-Smith 1954: 6):

> [PROBLEM 2–25]
> If 3 cats kill 3 rats in 3 minutes, how long will it take 100 cats to kill 100 rats?

> [PROBLEM 2–26]
> If it takes 3 3/4 minutes to boil one egg, how long will it take to boil six eggs?

In [2–25] the solver might conclude that 100 cats will kill 100 rats in 100 minutes. However, if the problem is extracted from its pull, the solution becomes obvious: if 3 cats kill 3 rats in 3 minutes, 100 cats will kill 100 rats also in 3 minutes (performing the action together of course), and in general, n cats will kill n rats in 3 minutes. Similarly, in [2–26], if the problem is considered outside of its pull, it should be obvious that it will take 3 3/4 minutes to boil any number of eggs, depending of course on the size of the pot. Such problems force the student to consider the language in which a problem is cast more closely. More will be said about this aspect of PS in chapter 4.

3
PROBLEMS IN ALGEBRA

I would advise you Sir, to study algebra, if you are not already an adept in it: your head would be less muddy, and you will leave off tormenting your neighbours about paper and packthread, while we all live together in a world that is bursting with sin and sorrow.

Samuel Johnson (1709–1784)

Algebra is arithmetic with letters. The main difference between solving a problem with letters and doing so with numbers lies in thinking about the problem in general rather than in specific number terms. The Greek mathematician Diophantus (who lived in the 3rd century CE) was probably the first to make this distinction, earning him the title "father of algebra." The chief semiotic difference between arithmetical and algebraic competence consists, therefore, in knowing how to represent a problem in particular versus how to represent a problem in general.

The history of algebra began, actually, in ancient Egypt and Babylon. The Babylonians solved equations by essentially the same procedures we use today. This ancient knowledge found a home early in the Arab world, where it was known as the "science of restoration and balancing." Indeed, the Arabic word for restoration, *al-jabr,* is the root of the word *algebra.* As mentioned in the first chapter, it comes from the title of a book written in the ninth century by Mohammed ibn-Musa al-Khwarizmi. In the mid-1100s, a Latin translation of al-Khwarizmi's book introduced the Hindu-Arabic numeral system to Europe. However, the system was largely ignored until Fibonacci's *Liber Abaci* was published (as discussed in chapter 1). Descartes later developed and standardized algebraic notational conventions—conventions that we use to this day.

This chapter will look at problems in algebra. There is nothing mysterious or arcane about using letters or symbols rather than actual numbers, as many students might initially think. Symbol-use is a fundamental semiotic representational strategy, as we have seen in previous chapters. In this case, letters are used to stand for "any number" or "a number in general." To perform algebraic operations efficiently and systematically new names are required. For example in $n^2 + 10$, the n stands for any number, known as a *variable* or an *unknown*; the 10 is called a *constant*; and the whole symbol is called an *algebraic expression*. The parts of an expression are called *terms*. Thus, $n^2 + 10$ is an algebraic expression made up of two terms, a variable (n^2) and a constant (10).

The technique of unraveling the value of unknowns is called *solving an equation*. The equation is a linear text representing two expressions as being equal in value to each other, as if they were on a balance scale:

$x + 3 = 5$ ← *Equation with one unknown*, x

An equation thus has two parts—one to the left of the equality sign (x + 3), and the other to the right (5). *Solving* the equation means finding out the value of the unknown, x, that will maintain the balance (in value terms):

> What number value does x have such that when 3 is added to it the result is 5?

The only number that answers this question is, of course, 2:

$$x + 3 = 5$$
$$\downarrow$$
$$2 + 3 = 5$$

This kind of reasoning is the essence of algebraic competence. It should be mentioned from the outset that the use of real-life analogues, such as the balance scale, helps students grasp new algebraic notions and techniques much more effectively (Quinlan 1995). So, again, concretization, annotation, contextualization, exemplification, etc. will form the basis of a semiotic pedagogy in this area of math PS.

Defining an Algebraic Problem

Consider the following problem:

> [PROBLEM 3–1]
> Blake is a growing child. If six years are added to his present age, his age will become double. How old is Blake at present?

The problem states that by adding 6 years to Blake's present age, it will be doubled. First, we must answer the question: What algebraic expression would stand for "Blake's age plus six years"? We start by letting a letter, say x, represent Blake's present age. This is the unknown. Now, by adding 6 to it we get the required expression:

Blake's age	*Plus*	*six years*
↓	↓	↓
x	+	6

Next, what kind of algebraic expression translates "double Blake's age"? We know that Blake's present age is x. Doubling his age is equivalent to multiplying his present age by 2:

x ← *Blake's present age*
2x ← *Twice his age*

Now, we can equate the two algebraic expressions:

$$x + 6 = 2x$$

To find out what x stands for (namely, to find out Blake's age), all we have to do is solve the equation. Solving equations will be discussed below. For now, it is sufficient to note that a large portion of algebraic PS is all about setting up equations, like this one. Equations are textual modeling devices, which convert a problem stated in ordinary language into algebraic language (Sebeok and Danesi 2000). As the renowned mathematician and educator George Polya (1887–1985) noted in his classic book on problem-solving, *How to Solve It* (Polya 1957: 174), the process of finding solutions to algebraic problems constitutes essentially "translation from one language into another." The following problem in plane analytic geometry was used by Polya (1957: 176–177) to illustrate this:

> Being given the equation of a straight line and the co-ordinates of a point, find the point which is symmetrical to the given point with respect to the given straight line.

The solution consists in translating the language of the problem to algebraic language. It is required to find a point, P_1 (p, q), which is symmetrical to a given point, P_2 (a, b), with respect to a given straight line. The general equation for a straight line is $y = mx + n$. The geometrical condition is that P_1 and P_2 are symmetrical to each other with respect to y. We note that "symmetrical" indicates the following geometrical relationships: (1) P_1 and P_2 are perpendicular to the given line; and (2) the midpoint of the line joining these points lies on the given line. The remainder of the problem involves setting up the relevant equation. As Polya (1957: 174) states:

> To set up equations means to express in mathematical symbols a condition stated in words; it is translation from ordinary language into the language of mathematical formulas. The difficulties which we may have in setting up equations are difficulties of translation.

In algebraic PS we must start by teaching the appropriate grammar and lexicon of the target language—algebra. As mentioned, a variable or unknown is an unspecified number, usually represented with a letter of the alphabet. An algebraic expression is, as we saw, an expression made up of variables and constants, connected to each other by plus or minus signs:

$$5x^2 + x - 12$$

This expression has three terms: $5x^2$, x, and -12. It is called a *trinomial* expression for this reason. An expression with one term is called a *monomial;* one with two a *binomial;* and so on. In general, an expression with an unspecified number of terms is called a *polynomial.* A variable can have a *coefficient*—if it is a number it is called a *numerical coefficient;* if it is another variable, it is called a *literal coefficient.* In $5x^2$ the digit 5 is the numerical coefficient of x^2; in ax^2, the variable a is the literal coefficient of the same variable. In all algebraic expressions, when no numerical coefficient is shown, it is assumed to be 1 or -1, as the case may be:

- In $x^2 + 1$ the numerical coefficient of x^2 is 1, and thus the expression is really $1x^2 + 1$.

- In $5x^2 - x$ the numerical coefficient of x is -1, and thus the expression is really $5x^2 - 1x$.

At this point, appropriate problems should be constructed that will get students accustomed to the new terminology and to get them to perceive algebra as a language. Here are two examples that are self-explanatory:

[PROBLEM 3–2]
Name all the terms and their parts in the following expressions. Then, indicate the type of expression each one is:
(1) $3x^2$
(2) $2x + 10$
(3) $(y - 1) / 4$

(4) 1.14n
(5) $2x^2 + 9x - 8$

[PROBLEM 3–3]

Translate the following statements into algebraic expressions:

(1) The square of a number is multiplied by 5 and 15 added to the result.
(2) A salary is increased by 1.
(3) In a group of 5 people, one person gets a gallon of syrup, and the remaining 4 get an equal amount of syrup.

Teaching the Algebraic Operations

Teaching the algebraic operations involves imparting the same notions and using the same kinds of techniques as teaching the arithmetical operations. Let's start with factoring and simplifying. Factoring can be carried out if the variables in an expression are of the same kind—all x's, all x^2's, etc. This is known as the *degree* of the variable, which is represented by an exponent. The variable x is really x^1 and is thus known as being of the first degree; x^2 is of the second degree; x^3 of the third degree; x^4 of the fourth degree and so on.

Basically, to factor an expression, all we have to do is "remove" the same-degree variable from its coefficients; and to simplify all we have to do is combine the remaining coefficients. Using concretization and annotation, this can be presented as follows:

Variables of the first degree:

$3x + 7x$
↓
$(3 + 7)\,x$ ← *Factoring = Removing the variable from its coefficients*
↓
$10x$ ← *Simplifying*

Variables of the second degree:

$x^2 - 4x^2 + 7x^2$
↓

$(1 - 4 + 7)\,x^2$ ← *Factoring = Removing the variable from its coefficients*

↓

$4x^2$ ← *Simplifying*

These two examples show that factoring and simplification can be reduced to removing and combining the coefficients of a variable. This is, of course, a process akin to morphological analysis in semiotics and linguistics (as mentioned in chapter 1). A word such as *incredibility* is composed of three morphemes: *in* + *credible* + *ty*. This kind of analysis (intuitive or reflective), by the way, allows us to identify a morpheme when it occurs in other forms or structures. For example, the *in-* morpheme is found in words such as *invisible, indelible,* and the like. Factorization in algebra is similarly a morphological procedure.

$$3x + 7x = (7 + 3)x = 10x$$
$$x^2 - 4x^2 + 7x^2 = (1 - 4 + 7)x^2 = 4x^2$$
$$7x^3 - 5x^3 = (7 - 5)x^3 = 2x^3$$
$$5x^4 - 2x^4 = (5 - 2)x^4 = 3x^4$$

If an expression contains variables of different degrees, we apply the same factorization procedure to those of the same degree separately. Here's an example:

$$4xy + 2x^2 - xy - 3x^2$$

First, we rearrange the similar terms as follows:

$$4xy - xy + 2x^2 - 3x^2$$

And then we factor them accordingly:

$$(4 - 1)\,xy + (2 - 3)x^2 = 3xy - x^2$$

Note that the resulting expression can be factored further, since x is a common factor in both its terms:

$$3xy - x^2 = x(3y - x)$$

To teach complex expressions, we first recall the rule of aggregates, reminding students that simplification starts with the innermost aggregate in parentheses (chapter 2). Here is an example:

$$\{5a + 6a + [5a - a + (3a + 4a)] - a\}$$

First, we simplify the expression in parentheses:

$$(3a + 4a) = 7a$$

We put this into the initial expression:

$$\{5a + 6a + [5a - a + 7a] - a\}$$

Next, we simplify the expression enclosed in square brackets:

$$[5a - a + 7a] = 11a$$

We plug this value in:

$$\{5a + 6a + 11a - a\}$$

Finally, we simplify the remaining terms:

$$\{5a + 6a + 11a - a\} = 21a$$

Now, let's look at other operations. To add expressions, we inform students that we simply add their like terms—terms of the same kind and degree—using the rules of arithmetic. For example:

$$(4x + 3y + 13) + (5x - 8y + 8) = ?$$

We can align the like terms under each other as in digit arithmetic and then add the separate columns in the usual fashion:

Aligning like terms:

$$\begin{array}{r} 4x + 3y + 13 \\ \underline{+5x - 8y + 8} \\ 9x - 5y + 21 \end{array}$$

Or else we can show students how to combine the like terms by removing the parentheses:

$(4x + 3y + 13) + (5x - 8y + 8)$		
$4x + 3y + 13 + 5x - 8y + 8$	←	*Removing the parentheses*
$(4x + 5x) + (3y - 8y) + (13 + 8)$	←	*Combining like terms*
$9x - 5y + 21$	←	*Simplifying*

The same procedures apply to subtraction. We must recall, however, the rule that a negative sign in front of an aggregate means that we must change the signs of all the terms in the expression when the parentheses are removed (chapter 2):

$(4x^2 + 3y + x - 13) - (2x^2 - 3y + 8)$		
$4x^2 + 3y + x - 13 - 2x^2 + 3y - 8$	←	*Removing the parentheses*
$(4x^2 - 2x^2) + (3y + 3y) + x - (13 + 8)$	←	*Combining like terms*
$2x^2 + 6y + x - 21$	←	*Simplifying*

Multiplication and division are treated in the same way as they are in arithmetic. A term, variable, or constant outside an aggregate multiplies each term within the aggregate. Here are a few examples:

$$x\,(3x + 4y) = (x \times 3x) + (x \times 4y) = 3x^2 + 4xy$$
$$3x\,(2x + 3y) = (3x \times 2x) + (3x \times 3y) = 6x^2 + 9xy$$

To divide expressions, we factor the expressions first (if at all possible). Take, for example:

$$\frac{7x^2 + 14x}{2x + 4} = ?$$

In the numerator, we can see that 7x is common to both terms. So, we can factor it out:

$$\frac{7x^2 + 14x}{2x + 4} = \frac{7x\,(x + 2)}{2x + 4} \leftarrow \textit{Factoring out the } 7x$$

In the denominator 2 is a common factor. So, we can factor it out:

$$\frac{7x\,(x + 2)}{2(x + 2)} = \frac{7x\,(x + 2)}{2(x + 2)} \leftarrow \textit{Factoring out the } 2$$

We can now see that (x + 2) is a common factor in the numerator and denominator and can thus be crossed out:

$$\frac{7x\,\cancel{(x + 2)}}{2\cancel{(x + 2)}} = \frac{7x}{2}$$

We remind students that division of the two expressions is possible if $x \neq -2$. If it does equal -2 then the denominator is 0, and division by 0 is not allowed:

$$2(x + 2) = 2\,(-2 + 2) = (2)\,(0) = 0$$

Sometimes it is useful to remove the parentheses from an expression without simplifying it. For example, the expression a + (b + c) can be rewritten as a + b + c, and a + (- b - c) as a - b - c. But if a minus sign precedes, then we must apply the usual rule: a - (-b - c) becomes a + b + c

We can now pass on to teaching an important operation—the multiplication of binomials::

$$(2n + 1)\,(3n + 5) = ?$$

This can now be taught with the "linear approach" exemplified above. We start by multiplying the first two terms in each expression together:

$$(2n + 1)\,(3n + 5) = 6n^2 \ldots$$

Next, we multiply the inner terms and outer terms, adding the resulting two products to the previous result:

$$(2n+ 1)(3n + 5) = 6n^2 + (3n + 10n) \ldots \rightarrow \textit{Simplifying} = 6n^2 + 13n \ldots$$

Finally, we multiply the last two terms together, adding the product to the previous result:

$$(2n + 1)(3n + 5) = 6n^2 + 13n + 5$$

At this point we can apply the same step-by-step procedure to the multiplication of (a + b) (a - b) to show that its is $a^2 - b^2$. This is an important form in algebra. We start by multiplying the first two terms in each expression:

$$(a + b)(a - b) = a^2 \ldots$$

Next, we multiply the inner terms and outer terms, adding the two products to the previous result:

$$(a + b)(a - b) = a^2 + (ab - ab) \ldots = a^2 + 0 = a^2\ldots$$

We then multiply the last two terms, adding the product to the previous result:

$$(a + b)(a - b) = a^2 + (- b^2) = a^2 - b^2$$

We can now use this fact to show how to divide expressions such as the following:

$$\frac{x^2 - 9}{4x - 12} = ?$$

First, we factor the numerator, noting that $(x^2 - 9)$ has the form $(a^2 - b^2)$:

$$\frac{x^2 - 9}{4x - 12} = \frac{(x + 3)(x - 3)}{4x - 12}$$

We note that 4 is a common factor in the denominator. So, we can factor it out:

$$\frac{(x + 3)(x - 3)}{4x - 12} = \frac{(x + 3)\,(x - 3)}{4(x - 3)}$$

The term (x - 3) is a factor in both the numerator and the denominator, and can thus be crossed out to produce a simplified form:

$$\frac{(x + 3)\,\cancel{(x - 3)}}{4\,\cancel{(x - 3)}} = \frac{(x + 3)}{4}\ (x \neq 3)$$

Needless to say, once the structural properties of these operations have been dissected and explained, it becomes a simple task to devise exercises based on them. These are, of course, similar to the traditional ones that math teachers have been using since time immemorial. The difference is that they can now be seen to put students in a mental position to see them as exemplifications of general structural properties of arithmetic, and thus not as exceptional in any way. The idea in semiotic pedagogy is, arguably, to instill a "meta-mathematical" mindset in students, so that they can see the interconnectedness of mathematical domains. Here are a few examples:

[PROBLEM 3–4]
Factor the following expressions, explaining the procedures used:
(1) $4x + 16$
(2) $15y^2 - 9y$

[PROBLEM 3–5]
Evaluate the following:
(1) $45x + 56x - 67x$
(2) $x(55x^2 + 2x)$
(3) $(x + 2)\,(3x - 7)$
(4) $(3x + 5y) - (6x - 4y - 7)$
(5) $(5x^2 - 2x - 6) \div (x^2 + 3)$
(6) $3x\,(x^2 - 4y)$

The topic of exponents can now be introduced. There are eight basic rules of exponents. Let's look at each one separately. The first one is as follows:

Rule of Exponents 1: $(x^a)(x^b) = x^{a+b}$

We point out that this rule pertains to identical bases and exemplify it with concrete examples, such as the following:

$$(5^3)(5^4) = 5^{3+4} = 5^7 = 78{,}125$$
$$(6^2)(6^3) = 6^{2+3} = 6^5 = 7{,}776$$

To help the learner see why this is so, we recall the discussion in the previous chapter. Using a schematic strategy students can be shown the relations as follows:

$$\begin{array}{ccc} 5^3 & \times & 5^4 \\ \downarrow & & \downarrow \\ (5 \times 5 \times 5) & \times & (5 \times 5 \times 5 \times 5) \end{array}$$

The exponents tell us, in effect, how many factors are involved in the multiplication, which is equivalent to adding up the number of exponents (4 + 5 = 9). Rule 1 simply generalizes this fact. A few other examples can be used to show that the rule applies to the multiplication of any number of bases (not just two) and to signed exponents as well:

$$\begin{array}{lclcl} (5^3)(5^4)(5^9)(5^0) & = & 5^{3+4+9+0} & = & 5^{16} \\ (6^{30})(6^4)(6^{-8})(6^{-1}) & = & 6^{30+4-8-1} & = & 6^{25} \\ \ldots & & & & \\ (x^a)(x^b)\ldots(x^n) & = & x^{a+b+\ldots+n} & & \end{array}$$

The second rule of exponents is as follows:

Rule of Exponents 2: $(x^a)(y^a) = (xy)^a$

In this case the exponents are the same while the bases are different. Here are two examples that concretize this rule:

$(5^3)\,(4^3) = (5 \times 4)^3 = 20^3 = 8{,}000$
$(6^2)\,(7^2) = (6 \times 7)^2 = 42^2 = 1{,}764$

To help the learner see why this rule is so, we can elaborate on what the first example really means, using a schematic strategy:

$$\begin{array}{ccc} 5^3 & \times & 4^3 \\ \downarrow & & \downarrow \\ (5 \times 5 \times 5) & \times & (4 \times 4 \times 4) \end{array}$$

The rule informs us, basically, to multiply these factors in combinations—the first 5 by the first 4, the second 5 by the second 4, and the third 5 by the third. This can be shown schematically as follows:

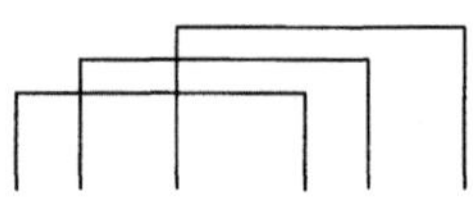

$$(5 \times 5 \times 5) \times (4 \times 4 \times 4) \quad = \quad (5 \times 4)\,(5 \times 4)\,(5 \times 4)$$

And this result is, of course, equivalent to $(5 \times 4)^3$. Again, Rule 2 applies to the multiplication of any number of bases (not just two) and to signed exponents as well:

$(5^3)\,(6^3)\,(7^3)\,(8^3) \quad = \quad (5 \times 6 \times 7 \times 8)^3 = \quad (1680)^3$
$(6^{-2})\,(-5^{-2})\,(1^{-2}) \quad = \quad (6 \times -5 \times 1)^{-2} \quad = \quad (-30)^{-2}$
…
$(x^a)\,(y^a)\ldots(z^a) \quad = \quad (xy\ldots z)^a$

The third rule is the following:

Rule of Exponents 3: $x^a/x^b = x^{a-b}\ (x \neq 0)$

We point out that this rule concerns identical bases, exemplifying it concretely:

$(5^4)/(5^3) \quad = 5^{4-3} \quad = 5^1 \quad = 5$
$(6^9)/(6^{-3}) \quad = 6^{9-(-3)} \quad = 6^{9+3} \quad = 6^{12}$

To help students see why this is so concretely, we inform them that the rule simply reflects what is involved in division:

$$\frac{5^4 = \cancel{(5 \times 5 \times 5)} \times 5}{5^3 = \cancel{(5 \times 5 \times 5)}} = 5$$

The exponents tell us how many factors are involved in the division. As can be seen, the end result is equivalent to subtracting the number of exponents (4- 3 = 1). The rule simply generalizes this fact.

The fourth rule of exponents is as follows:

Rule of Exponents 4: $\left(\frac{x}{y}\right)^a = \frac{x^a}{y^a} \qquad (y \neq 0)$

We point out that this rule applies to different (fraction) bases with identical exponents. The following concrete examples show this clearly:

$$\left(\frac{4}{5}\right)^3 = \frac{4^3}{5^3} = \frac{64}{125}$$

$$\left(\frac{7}{9}\right)^5 = \frac{7^5}{9^5} = \frac{16{,}807}{59{,}049}$$

The reason why this is so can be shown easily by telling students to consider what the first example means:

$$\left(\frac{4}{5}\right)^3 = \left(\frac{4}{5}\right)\left(\frac{4}{5}\right)\left(\frac{4}{5}\right) = \frac{4^1 \times 4^1 \times 4^1}{5^1 \times 5^1 \times 5^1} = \frac{4^{1+1+1}}{5^{1+1+1}} = \frac{4^3}{5^3}$$

The fifth rule is as follows:

Rule of Exponents 5: $(x^a)^b = x^{ab}$

Here are two specific uses of this rule:

$(2^3)^4 = 2^{(3)(4)} = 2^{12}$
$(3^{-3})^5 = 3^{(-3)(5)} = 3^{-15}$

To help students see why this is so, we tell them to consider what the first example means:

$$(2^3)^4 = (2^3)\,(2^3)\,(2^3)\,(2^3) = \textit{(Using Rule 1 above)}\ 2^{3+3+3+3} = 2^{(3)(4)} = 2^{12}$$

The sixth rule is as follows:

Rule of Exponents 6: $x^0 = 1\ (x \neq 0)$

To help the student see why this is so, we can work through a concrete example, such as 5^0. The 0 exponent results from any division in which the numerator and the denominator have the same number of factors. An example is the following:

$$\frac{5^5}{5^5} = \frac{(5 \times 5 \times 5 \times 5 \times 5)}{(5 \times 5 \times 5 \times 5 \times 5)} = \frac{\cancel{(5 \times 5 \times 5 \times 5 \times 5)}}{\cancel{(5 \times 5 \times 5 \times 5 \times 5)}} = 1$$

From Rule 3 we know that $5^5/5^5$ is equal to 5^{5-5} or 5^0. Therefore, $5^0 = 1$. Again, we emphasize that Rule 6 is a formalization of this fact.

The seventh rule is shown below:

Rule of Exponents 7: $x^{-a} = 1/x^a$

This formalizes a structural fact pertaining to reciprocals. A *reciprocal* is a number related to another in such a way that when multiplied together their product is 1. For example, the reciprocal of 7 is 1/7 (and $7 \times 1/7 = 1$) ; the reciprocal of 2/3 is 3/2 (and $2/3 \times 3/2 = 1$); and so on. Here are two examples of the above rule:

$5^{-3} = 1/5^3$
$6^{-13} = 1/6^{13}$

To get the student to see why this is so, we recall from Rule 6 that $x^0 = 1$. So, let's substitute this in place of 1 in the numerator of $1/x^a$:

$$\frac{1}{x^a} = \frac{x^0}{x^a}$$

Now, we know from Rule 3 that this is equivalent to x^{0-a} which is, of course, equal to x^{-a}. This proves that $x^{-a} = 1/x^a$.

The eight and final rule of exponents is as follows:

Rule of Exponents 8: $x^{1/a} = {}^a\sqrt{x}$

Here are two examples:

$$5^{1/2} = \sqrt{5}$$
$$6^{1/3} = {}^3\sqrt{6}$$

To show why this is so, all we have to do is take $5^{1/2}$ and square it. The result is 5:

$(5^{1/2})^2 = 5^{(1/2)(2)}$ *(Using Rule 5)* $= 5$

Since we get the same result by squaring $\sqrt{5}$, we have proved that $5^{1/2} = \sqrt{5}$. At this point of the pedagogical process, it is essential to get students to work with the new ideas through PS. As has been discussed (or at least implied) throughout this and previous chapters, PS, representation, and discovery in math are intrinsically interrelated. One implies the other synergistically. The history of mathematics is characterized by problems or notational creations (such as the invention of exponents) that have led serendipitously to mathematical discoveries. The reason for this is that, as Dehaene (1997: 151) aptly puts it, when mathematicians contemplate a certain problem, "truth descends on them," leading them to imagine some new way of doing something and thus opening up new vistas and horizons. Imparting a comparable feeling of discovery, albeit to a limited and controlled extent, is what semiotic pedagogy is all about, since it is based on interrelating notation (representation), PS, and understanding of math concepts.

The following types of problems can be constructed to get students to work concretely with the above rules. Again, these are not exceptional, in that they are found in most pedagogical approaches to algebraic PS. The semiotic goal is to link algebra to arithmetic and, thus, impart a metamathematical awareness in students of their interrelatedness. They are self-explanatory and thus require no particular commentary:

[PROBLEM 3–6]

Evaluate the following, explaining what you are doing and why:

(1) 6^{-4}

(2) $(3^{3})(3^{7})(3^{-5})$

(3) $(10^{4})(9^{4})$

(4) $7^{14} \div 7^{9}$

(5) $(5/6)^{4}$

(6) $(5^{2})^{3}$

(7) $3{,}400^{0}$

(8) $27^{1/3}$

[PROBLEM 3–7]

Now evaluate the following, similarly explaining what you are doing and why:

(1) $(x^{3})(x^{19})$

(2) $(x^{3})(y^{3})$

(3) $(xy)^{4} \div y^{4}$

(4) $(x + a)^{3}(x + a)^{9}$

(5) $(x^{-13})(y^{-13})$

(6) $(x^{1/3})(x^{1/3})$

Teaching Equations

Recall the equation that we set up above to represent Blake's age (problem [3–1]):

$$x + 6 = 2x$$

To teach the student how to solve the equation, the real-life concept of a balance scale can be used pedagogically, since the two sides of the equation can be compared to the two pans on a scale, whereby the value of the terms on the left side must equal the value of the terms on the right side. Using this concretization technique, we can now formalize equation-solving. First, we indicate that we must subtract the x on the left side. This means that we must subtract it as well from the other side, to keep the balance intact:

$$(x - x) + 6 = 2x - x$$

Now, $(x - x)$ on the left side gives, of course, the result of 0. The effect of subtracting x from both sides has thus been to eliminate the x from the left side of the original equation:

$$6 = 2x - x$$

Simplifying the expression on the right side gives the result of 1x or simply x. This allows us to rewrite the equation as:

$6 = x$

$x = 6$ ← *Turning the sides around*

This is, in effect, the solution, namely that $x = 6$. It tells us that Blake is 6 years old. To check if this is, indeed, the correct solution, all we have to do is use this information in the original problem. Indeed, if we add 6 years to Blake's 6 years we get $6 + 6 = 12$ years, which is equivalent to doubling his age, $6 \times 2 = 12$. Story problems (such as this one) will be discussed in the next chapter. Suffice it to say here that they are critical PS texts for formalizing to students what equations are all about. Relevant studies have found that even children can solve equations when they are cast in story problem forms, because these are perceived to have real-life meaning (Liebeck 1984, Musser, Burger, and Peterson 2006). I have noticed in my own classes that learners intuitively grasp the concept of equations and how to set them up if the latter is explained to them as a sign-substitutive process (Radford and Grenier 1996). A study by Swafford and Langrall (2000), for instance, showed that even sixth-grade students with no formal training in algebra could easily solve specific kinds of problems and generalize the problem situation using variables and equations.

Once the appropriate structural concept of equations has been imparted, a shortcut method for solving them can be introduced, explaining it on the basis of how the above solution process unfolded. The end result of subtracting x from both sides was to "move the x on the left side over to the right side" changing its sign in the process:

$x + 6 = 2x$

$6 = 2x - x$ ← *Moving* x *over to the right side, changing its sign*

$6 = x$ ← *Simplifying*

$x = 6$ ← *Rearranging*

This procedure will thus allow us to skip the step that eliminates a term on either side. It suggests a general procedure for *transposing* terms from side to side, as it is called: that is, we can eliminate something from any side of the equation by "bringing it over" to the other side, changing its sign in the process. For example:

$2x + 5 = 9$

$2x = 9 - 5$ ← *Transposing 5 and changing its sign*

$2x = 4$

$x = 2$

$3x + 7 = 9 - x$

$3x + x + 7 = 9$ ← *Transposing -x and changing its sign*

$4x + 7 = 9$

$4x = 9 - 7$ ← *Transposing 7 and changing its sign*

$4x = 2$

$x = 2/4$ *or* $1/2$

At this point, we should name this type of equation as *linear*. We can also explain that it is so named because, if plotted on graph paper, the equation produces a line. We can also point out that equations of the first degree are all linear.

In the solutions above a number of things were taken for granted. So, it is always best to go through each step concretely. Let's take the following as a case-in-point:

$3x - 4 = 8$

The objective in solving equations is, as we have seen, to get the variable x to "stand alone" on the left side with no coefficients or other terms. So, this means eliminating the -4 and the coefficient 3 (in 3x). Let's start by eliminating the -4, using the general procedure of transposing it to the right side while changing its sign from -4 to +4:

$3x = 8 + 4$

We remind the student that this is a shortcut equivalent to adding 4 to both sides. Next, we simplify the right side:

$$3x = 12$$

Now, how do we eliminate the coefficient 3? Dividing 3x by 3 would do the job, since 3x/3 allows us to cancel out the 3's from the numerator and denominator, leaving x. However, if we do this to the left side we must do the same to the right side, reminding the student of the balance schema:

$$3x \div 3 = 12 \div 3$$

Simplifying both sides we get:

$$x = 4$$

At this point, we can now go on to teach the fact that linear equations may have more than one variable. To solve linear equations with two variables, x and y, requires two equations. The two equations constitute a *system* of simultaneous equations. Here's an example:

(1) $4x + 3y = 13$
(2) $x + 2y = 2$

There are several ways to solve such a system. The most common way, known as the *substitution* method, is to express one variable in terms of the other and then substitute it into the other equation. How do we do this, the student might ask? We tell the student to look at equation (2). By transposing the 2y over to the right side, we get:

$$x = 2 - 2y$$

We now can plug in this value, (2 - 2y), for x in equation (1) above.

$$4x + 3y = 13$$

$$\downarrow$$

$$4(2 - 2y) + 3y = 13$$

Now we can solve for y in the usual fashion:

$8 - 8y + 3y = 13$ ← *Carrying out the required operations*

$-8y + 3y = 13 - 8$	← *Transposing 8 and changing its sign*
$-8y + 3y = 5$	← *Simplifying the right side*
$-5y = 5$	← *Simplifying the left side*
$y = -1$	← *Dividing both sides by -5*

Since we have discovered that $y = -1$, we can put its value in equation (2) above:

$x + 2y = 2$	← *Equation (2)*
$x + 2(-1) = 2$	← *Plugging in -1 for y*
$x - 2 = 2$	← *Simplifying the left side*
$x = 2 + 2$	← *Transposing -2 and changing its sign*
$x = 4$	← *Simplifying the right side*

We now have uncovered the value for x as well. The solution is, thus: $x = 4$, $y = -1$. A second method for solving systems of equations is known as the *elimination method* or the *Gauss method*. The objective is to make the coefficients of one variable the same in both equations, so that the variable can be eliminated by either subtracting or adding the equations. Clearly, the schema of an equilibrium device, in a different form, is again pedagogically propitious. Take the same equations:

(1) $4x + 3y = 13$
(2) $x + 2y = 2$

If the coefficient of x in equation (2) were 4, then we could subtract (2) from (1) to eliminate the x variable. So, let's go ahead and multiply equation (2) by 4:

$4(x + 2y) = (2)(4)$	← *Multiplying both sides of equation (2) by* 4
$4x + 8y = 8$	← *Simplifying both sides*

We can now put this equivalent equation under (1):

(1) $4x + 3y = 13$
(2) $4x + 8y = 8$

Let's subtract the two equations:

(1) $4x + 3y = 13$
(2) $\underline{-(4x + 8y) = -8}$

(1) $4x + 3y = 13$
(2) $\underline{-4x - 8y = -8}$ ← *Removing the parentheses with sign changes*
$-5y = 5$ ← *Subtracting (2) from (1)*
$y = -1$ ← *Dividing both sides by -5*

Replacing y with -1 in either of the original equations allows us to determine that x = 4. We can now plug it into equation (1):

$4x + 3y = 13$
$4x + 3(-1) = 13$ ← *Plugging in -1 for y*
$4x - 3 = 13$ ← *Simplifying the left side*
$4x = 13 + 3$ ← *Transposing the -3 and changing its sign*
$4x = 16$ ← *Simplifying the right side*
$x = 4$ ← *Dividing both sides by 4*

Having presented the concepts and shortcut procedures above, we can now discuss the equation known as *Diophantine*, defined as a single linear equation with two variables. The solution in this case involves open-ended PS, since informed guessing to solve for both unknowns will be required. Because of this feature of Diophantine equations, they should not be introduced earlier in the pedagogical sequence. As an open structure it involves exploration based on previous know-how. For example, what possible values could x and y have if we are given the following single equation?

$2x + 3y = 20$

Solving such equations involves exploratory trial and error. Its possible solutions are charted below:

Value for x	*2x*	*Value for y*	*3y*	*Result*
1	2	6	18	2 + 18 = 20
4	8	4	12	8 + 12 = 20
7	14	2	6	14 + 6 = 20

Any value higher than $x = 7$ would not work. For instance, if 8 is assigned to x, then $2x = 16$. This means that 3y should equal 4 in order for the equation to have the value of 20. However, there is no integral value that can be assigned to y to make $3y = 4$. If 9 is assigned to x, then $2x = 18$. But then any value assigned to y in 3y would make the equation greater than 20 in value, since the lowest value for y is 1, and $3y = 3$. Adding these would produce 21, not 20 as required by the original equation. Note that in the chart the second solution set is also ruled out because x and y both equal 4.

In an elementary or introductory course, the topic of quadratic equations comes up. The technique of opposition comes in handy once again, since a quadratic equation can be seen to form a binary opposition with a linear equation. The latter is, of course, the unmarked form because it contains the lowest (and thus most basic) degree of the variable. So, we start by reminding students that a linear equation is one in which the degree of the variable is 1:

$$x^1 + 4 = 6 \leftrightarrow x + 4 = 6$$

A *quadratic equation* (from "square" in Latin) is one in which the degree of the variable is 2:

$$(1)\ x^2 + 4 = 20$$
$$(2)\ 3x^2 + 4 + 2x = 20$$

A quadratic equation, more accurately, is an equation that has both second- and first-degree variables in it, as can be seen overtly in (2). The first-degree variable in (1) is equal to 0 and is thus not shown. Solving a quadratic equation involves the same types of procedures used to solve linear equations. But before illustrating and discussing a solution, it is essential to put quadratic equations into their prototypical form. In the case of (2), this can be done as follows:

$3x^2 + 4 + 2x = 20$ ← *Quadratic equation*
$3x^2 + (4 - 20) + 2x = 0$ ← *Transposing 20 and changing its sign*
$3x^2 -16 + 2x = 0$ ← *Simplifying the left side*
$3x^2 + 2x - 16 = 0$ ← *Rearranging the terms on the left side*

This layout of (2) now fits the general structure of the quadratic equation, which is the following:

$$ax^2 + bx + c = 0$$

The a and b are the numerical coefficients of x^2 and x respectively. Thus, in equation (2), these would be a = 3 and b = 2.

$$\begin{array}{l} ax^2 + bx + c = 0 \\ \downarrow \quad \downarrow \\ 3x^2 + 2x - 16 = 0 \end{array}$$

The letter c represents the constant in the equation. In our equation it stands for -16:

$$\begin{array}{l} ax^2 + bx + c = 0 \\ \qquad\qquad \downarrow \\ 3x^2 + 2x - 16 = 0 \end{array}$$

The first equation above, $x^2 + 4 = 20$, can also be made to fit the general quadratic structure. As mentioned, in this case the coefficient of x is 0. For the sake of clarity, this can be added to the equation:

$x^2 + 0x + 4 = 20$
$x^2 + 0x + (4 - 20) = 0$ ← *Transposing 20 to the left side*
$x^2 + 0x - 16 = 0$ ← *Simplifying the left side*
$x^2 - 16 = 0$

This layout can now be mapped against the general quadratic form:

$$\begin{array}{l} ax^2 + bx + c = 0 \\ \downarrow \quad \downarrow \quad \downarrow \\ 1x^2 + 0x - 16 = 0 \end{array}$$

We can now proceed to teaching the strategies involved in solving quadratic equations. There are three methods for solving them. The first is to factor the equation, if this possible. Consider the first equation above (in its prototypical form):

$$x^2 - 16 = 0$$

Using the insights for factoring already taught, it can be seen that $x^2 - 16 = (x + 4)(x - 4)$. Plugging this in the left side of the equation above, we get:

$(x + 4)(x - 4) = 0$ ← *Factoring the left side*

Now, from arithmetic we know that if the product of two factors is 0, then one of the factors (or both) must be zero. So:

(1) $x + 4 = 0$
Or
(2) $x - 4 = 0$

Solving for x in both cases, we get:

(1) $x = -4$ ← *Transposing 4 and changing its sign*
(2) $x = +4$ ← *Transposing -4 and changing its sign*

We alert students to the fact that there are always two solutions to such equations: $x = +4$, $x = -4$, or simply, $x = \pm 4$ (± stands for "plus or minus"). Set notation can also be used: $x = \{+4, -4\}$. The differential signifiers in the linear-versus-quadratic opposition can now be summarized in chart form as follows:

	Linear Equation	*Quadratic Equation*
Variable	First degree (x)	Second degree (x^2)
Solution	One	Two (with opposite signs)

The second method is called *completing the square.* An expression such as the following is called a perfect square because it can be rewritten in the general form $(x + y)^2$:

$x^2 + 2xy + y^2 = 0$
$(x + y)^2 = 0$ ← *Because* $(x + y)^2 = (x + y)(x + y) = x^2 + 2xy + y^2$

Using this insight, we can now solve some quadratic equations, such as $x^2 + 8x + 16 = 0$. The latter equation can, in fact, be rewritten as $(x + 4)^2$. Now, what about equations that are not perfect squares, such as $x^2 + 8x + 15 = 0$? We start as follows:

$x^2 + 8x + 15 = 0$
$x^2 + 8x = -15$ ← *Transposing 15 and changing its sign*

We then add 16 to both sides of the equation, in order to produce a perfect square on the left side. This does not alter the values of the equation since the same number has been added to both sides:

$x^2 + 8x + 16 = -15 + 16$

$x^2 + 8x + 16 = 1$ ← *Simplifying the right side*

$(x + 4)^2 = 1$ ← *Factoring the left side*

$\sqrt{(x + 4)^2} = \sqrt{1}$ ← *Taking the square root of both sides*

Results

$(x + 4) = 1$

$x + 4 = 1$ ← *Removing the parenthesis*

$x = 1 - 4$ ← *Transposing the 4 and changing its sign*

$x = -3$ ← *Simplifying the right side*

Or

$-(x + 4) = 1$

$-x - 4 = 1$ ← *Removing the parenthesis*

$-x = 4 + 1$ ← *Transposing the –4 and changing its sign*

$-x = 5$ ← *Simplifying the right side*

$x = -5$ ← *Dividing both sides by -1*

Solution Set

$x = \{-3, -5\}$

This method is a bit difficult to use and thus, is best avoided initially, unless students can work with it right away. The third method for solving a quadratic equation is to use the following formula based on the general form: $ax^2 + bx + c = 0$:

$$\frac{-b \pm \sqrt{(b^2 - 4ac)}}{2a}$$

This looks daunting to many students. Thus concretization and annotation are crucial as pedagogical techniques in showing how it can be used. Let's apply it to the equation above:

$x^2 + 8x = -15$

First, we transpose the -15 to the left side so that the equation will conform to the structure of the general formula ($ax^2 + bx + c = 0$):

$x^2 + 8x + 15 = 0$

Now, we label each part of the equation in terms of the a, b, c parts in the formula. We note that the coefficient of x^2 is 1, even though it is not explicitly shown. Thus a = 1, b = 8, and c = 15. We plug these values in the formula, using aggregate signs to help us keep values distinct:

$$x = \frac{-b \pm \sqrt{(b^2 - 4ac)}}{2a}$$

$$x = \frac{-8 \pm \sqrt{\{[8]^2 - 4[(1)(15)]\}}}{(2)(1)}$$ ← *Plugging in* a = 1, b = 8, c = 15

$$x = \frac{-8 \pm \sqrt{(64 - 60)}}{2}$$ ← *Simplifying*

$$x = \frac{-8 \pm \sqrt{4}}{2}$$ ← *Simplifying further*

$$x = \frac{-8 \pm 2}{2}$$ ← *Simplifying further still*

The solution set is therefore:

$x = (-8 + 2)/2 = -6/2 = -3$
Or
$x = (-8 - 2)/2 = -10/2 = -5$

Solution set
$x = \{-3, -5\}$.

For students who may be interested in grasping how the formula was derived, we can point out it is the result of using the method of completing the square and solving for x in the general formula. Students can thus be assigned the derivation of the formula as a problem. Before leaving the topic of quadratic equations, let's take a brief but interesting historical side journey, a journey that allows us to introduce the concept of imaginary number. As mentioned, the use of historical anecdotes is not just an interesting bit of "motivational spice," it constitutes an important contextualization strategy. We start by asking the student to consider the solution to the following quadratic equation:

$$x^2 = -1$$

Taking the square root of both sides produces:

$$x = \sqrt{-1}$$

What kind of number is the square root of a negative number? As the student knows by now, squaring a negative number produces a positive number, never a negative one:

$$-2^2 = -2 \times -2 = +4$$
$$-3^2 = -3 \times -3 = +9$$
$$-4^2 = -4 \times -4 = +16$$
$$\ldots$$
$$-n^2 = +n^2$$

So, what kind of number is $\sqrt{-1}$? Not knowing what to call it, mathematicians referred to it, logically enough, as an *imaginary number*. This is yet another case of a serendipitous discovery. As a matter of fact, mathematicians have used it to amplify the concept of number system. An imaginary number is now considered to be part of what is known as a *complex number*, which is defined as a number having the form a + bi, where a and b are real numbers and i is the square root of -1. The name *imaginary* is due to Descartes in the seventeenth century and the use of the symbol $i = \sqrt{-1}$ is due to Euler in the eighteenth century. Incredibly, complex numbers turn out to have many applications. They are used, for instance, to describe electric circuits and electromagnetic radiation and they are fundamental to quantum theory in physics.

The discovery of such numbers is probably due to the great twelfth-century Hindu mathematician Bhaskara (1114–1185). In his treatise on equations, *Ars magna* ("The Great Art"), Italian mathematician Girolamo Cardano (1501–1576) introduced the concept of the square root of a negative number to a European audience. His compatriot Rafael Bombelli (1526–1572) considered complex numbers to be solutions of cubic equations—equations in which one of the variables is cubed (x^3). Bombelli was the first to provide a consistent theory of complex numbers in his *Algebra* of 1572, specifying how the four operations (addition, subtraction, multiplication, and division) could be performed with complex numbers. It was Jean-Robert Argand (1768–1822) who gave complex numbers their first true comprehensive treatment, followed by Carl Friedrich Gauss's exhaustive study of such numbers. Interestingly, every real number can be represented as a complex number, by simply letting the imaginary part equal 0. So, for example, 5 can be represented as follows:

$$a + bi = 5 + 0i = 5$$

There are many equations that have no real number solutions, such as $x^2 + 1 = 0$. However, if x is allowed to be complex, the equation has the solutions $x = \pm i$. For example, $(\sqrt{-4})(\sqrt{-9})$ does not equal $\sqrt{36}$ or 6. Rather, it equals $(2i)(3i) = 6i^2 = 6(-1) = -6$.

Constructing problems to get students to reinforce their knowledge of equations is now a simple task. The order of the problems should follow the order of presentation above, which was based on structural reasoning. The latter started with simple linear (one variable) equations progressing to quadratic equations (the marked form of the basic linear-versus-quadratic opposition). The problems are self-explanatory:

[PROBLEM 3–8]

Solve for x:

(1) $x + 3 = 7x - 3$

(2) $5x + 15 = 25$

(3) $6x + x - 12 = 36 + 2x - 88$

(4) $x - 3x = 57 + x$

[PROBLEM 3–9]

Solve the following systems of equations:

(1) $3x + 2y = 10$
$4x + 5y = 18$

(2) $2x - y = 30$

$x + y = 54$

[PROBLEM 3–10]

Solve the following quadratic equations by the factoring method:

(1) $2x^2 - x = 6$

(2) $x^2 + 8x = -15$

(3) $4x^2 - 9 = 0$

[PROBLEM 3–10]

Solve the same quadratic equations above with the general formula:

(1) $2x^2 - x = 6$

(2) $x^2 + 8x = -15$

(3) $4x^2 - 9 = 0$

[PROBLEM 3–11]

What kind of solution can you find for $x^2 = -2$?

Algebraic Competence

The main subtext that I have attempted to interweave in this chapter is that the development of algebraic competence entails knowing how to extend arithmetical concepts into general representational (or formulaic) ones. Simply put, algebra involves the semiotic ability to represent arithmetical structures and oppositions in general ways.

The study of algebra brings out that there are four sign-based levels or orders involved in learning how to go from counting to equation-solving. The first order is the instinctive ability itself to count. This is the referential domain of mathematics (at least in part). Using numerical signs to stand for counting constitutes a second order. This is a representational order and can be called, simply, the order of numeration. It is the level at which counting concepts are "semiotized," so to speak, into numerals. The third order is the arithmetical one. This is the level at which numerals are organized into a code of operations based on counting processes (adding, taking away, comparing, dividing, and so on). Finally, algebra constitutes a fourth-order language that has the capacity to generalize the features and patterns of arithmetic by simply representing them with signs called variables. Algebra is not a code, therefore, but a "metalanguage"—a strategy for symbolizing arithmetical structure in generic form:

Fourth Order	Algebra = developing a metalanguage for arithmetic
Third Order	Arithmetic = encoding counting processes into rules and oppositions
Second Order	Numeration = using numeral signs to stand for counting units
First Order	Counting = the instinctive ability to separate referents into units that can be compared to each other in terms of quantity

The emergence of algebraic competence in students is, more than likely, the end result of following the above "natural semiotic order," as it can be called, in the instructional sequence and in the preparation of appropriate PS materials. Actually, this has always been the implicit curricular order used in math education. All the semiotic argumentation does, in effect, is give traditional pedagogical practices a solid theoretical justification. Because algebraic competence is a fourth-order semiotic skill, it should always be kept in mind that it requires a considerable amount of time to develop in students. The goal of semiotic instruction is, in effect, to "go with the natural flow," so to speak. All the teacher can do is present notions and problems in a flow that goes from concrete knowledge of counting to highly abstract knowledge of how to represent counting structures, emphasizing to students that algebra is a metalanguage for symbolizing how we count in general ways.

Using Historical Anecdotes

In the teaching of algebraic PS, we should always stress to students that algebra is not just a convenient shorthand code for solving counting problems. Symbolism plays a critical role in how humans discover things. Contemplating how we count in general ways often leads to discoveries on its own, as we have already seen several times in this book. This point is a crucial one for students to grasp, since it contextualizes the learning process in powerful psychological ways. The connection between symbols and the reality they represent is a dynamic one, with one guiding the other. The learning-effectiveness of bringing out this dynamic in the classroom cannot be emphasized enough. One anecdote that is very useful in this regard, given its long and fascinating history, is Fibonacci's famous Rabbit Puzzle, which the practical-minded medieval mathematician created simply to illustrate the advantages of using the decimal numeration system to his fellow Italians. As it turns out, no other puzzle has had as many implications for the study of numerical pattern as has the Rabbit Puzzle; and no other puzzle has had as many "reifications"—manifestations in natural phenomena and human products—as has Fibonacci's ingenious problem (Adam 2004, Posamentier and Lehmann 2007). It is worthwhile going through the main outline of the story concerning this puzzle here to bring out, as a concluding reflection on the theme of this chapter, how interesting it can be in a classroom setting. I should mention that I have often used this puzzle in my own classroom to show how puzzle-solving has often guided mathematicians in the past to discover new ideas and how it will also guide students to discover underlying principles of mathematical structure. Semiotic pedagogy attempts, in effect, to bring out how mathematical representation, PS, and reality mirror each other in isomorphic ways. The Fibonacci Rabbit Puzzle is an historical anecdote well worth remembering and using pedagogically, in my view at least.

Incidentally, there is no textual evidence to suggest that Fibonacci himself was aware of the implications and applications that the solution to his puzzle would turn out to have. He may or may not have. We simply do not know. It was the French mathematician François Edouard Anatole Lucas (1842–1891) who noticed its far-reaching implications in the nineteenth century. Since Lucas's observations, the amount of mathematical properties that Fibonacci's puzzle has been found to conceal and the number of reifications that it has been found to have in Nature and human life have been absolutely astounding. The question that the puzzle begs is an obvious one: How could such a simple problem, designed originally to represent the efficiency of decimal numerals over Roman ones in keeping track of counting processes, contain so many "secrets of the universe," so to speak?

As Umberto Eco (1998) has convincingly argued with regard to discovery in general, the Fibonacci sequence (as the solution to the Rabbit Puzzle is called) constitutes one of those truly intriguing episodes in human history which show how serendipity and discovery are intertwined. It is an episode with enormous implications for the study of representation, since it suggests that it is hardly an arbitrary symbolic process, but one that seems to constitute an unconscious probe of reality leading, in and of itself, to the discovery of hidden principles that govern human life and the universe.

Signs give shape to formless ideas, not in an arbitrary fashion, but in response to our experience of reality. As Peirce (1931–1958) often pointed out in his writings, signs are simulations of reality and this is why they can easily substitute that reality in our mind. The organization of signs into codes (such as the arithmetic code) produces knowledge systems. These vary throughout the world, presumably because the experiences that signs simulate vary according to culture and historical era. But such variation is, upon closer scrutiny, superficial. All sign systems display a cognitive deep structure, as it can be called—that is, below the surface of knowledge systems are sign-making processes that reflect universal tendencies in the perception of reality. In effect, all sign systems are valid "theories" of reality, evaluating it in culturally- and historically-based ways. In other words, our knowledge systems can only give us partial glimpses of reality. What is important to note is that the signifying structures that constitute these systems are hardly the product of firm reasoning processes. Nor are they simply products of human intentionality. Rather they seem to come to consciousness serendipitously. Discoveries cannot be forced by logical analysis or intent. They simply happen. But discovery is not magical or random, as the Fibonacci Rabbit puzzle episode shows. It is probably tied to unconscious processes of understanding reality that humans convert into sign processes. The latter then function as mental maps for further exploration, leading to subsequent discoveries. This is arguably the reason why representation in one realm of knowledge leads, later, to discoveries in other realms.

The word *serendipity*, incidentally, was coined by Horace Walpole in 1754, from the title of the Persian fairy tale *The Three Princes of Serendip,* in which three princes make a discovery accidentally. The tale can be retold somewhat as follows. Three princes from Ceylon were journeying in a strange land when they came upon a man looking for his lost camel. The princes, who had never seen the animal, asked the man a series of seemingly pertinent questions:

Was it missing a tooth?
Was it blind in one eye?
Was it lame?
Was it laden with butter on one side and honey on the other?
Was it being ridden by a pregnant woman?

Incredibly, the answer to all their questions was yes. The camel owner instantly accused the princes of having stolen the animal since he concluded logically that the princes could not have had such precise knowledge otherwise. But the princes explained their questions as based on simple observations. They had noticed that the grass on either side of the road was uneven and that this was most likely the result of the camel being lame and blind in one eye. They had also noticed that parts of the grass were chewed unevenly, suggesting a gap in the camel's mouth. They then told the owner that the uneven patterns of footprints were signs of awkward mounting and dismounting, which could be related to uneven weights on the camel. Given the society of the era, this suggested that the camel was ridden by a pregnant woman, creating a lack of equilibrium and thus an uneven pattern of footprints. Finally, in noticing differing accumulations of ants and flies they concluded that the camel was laden with butter and honey—the natural attractors of these insects. Their questions were, as it turns out, inferences based on astute observations. To use Peircean terminology, they were "abductions" of a logico-inferential nature (rather than inductions or deductions). Himself a logician and mathematician, Peirce argued that induction and deduction by themselves hardly explain how mathematical discovery comes about. In mathematics hunches and guesses play a much more central role in the origination of new ideas than is often assumed. Peirce termed the inferential process used in the origination process abduction. He described it as follows (Peirce 1931–1958, V: 180):

> The abductive suggestion comes to us like a flash. It is an act of insight, although of extremely fallible insight. It is true that the different elements of the hypothesis were in our minds before; but it is the idea of putting together what we had never before dreamed of putting together which flashes the new suggestion before our contemplation.

Ceylon's ancient name was Serendip, and it was Walpole who, after having read the tale, decided to introduce the word *serendipity* into the English language. Walpole termed the princes' discovery of the truth the result of using an innate "accidental sagacity." Serendipity characterizes the history of discovery in mathematics and science—Wilhelm Conrad Roentgen (1845–1923) accidentally discovered X-rays by seeing their effects on photographic plates; Alexander Fleming (1881–1955) serendipitously discovered penicillin by noticing the effects of a mold on bacterial cultures; and the list could go on and on (Merton and Barber 2003). The history of science makes it obvious that discovery is a hardly the product of a systematic search for truth, but rather a serendipitous consequence of using our abductive brain.

What is perhaps even more astounding is the fact that serendipity plays a role in reification—the manifestation of a form in referential domains other than the original one in which it was forged. A perfect example of this are the reifications of π (pi) = 3.14 (Beckmann 1981, Blatner 1997, Posamentier 2004, Eymard, Lafon, Wilson, and Lafon 2004). Pi is, of course, the ratio that results when the circumference of a circle is divided by its diameter. Serendipitously, π is appears in a number of mathematical calculations and formulas, such as the one used to describe the motion of a pendulum or the vibration of a string. It also turns up in equations describing the DNA double helix, rainbows, ripples spreading from where a raindrop falls into water, waves, navigation problems, and on and on. Does this mean that the circle form is implicit in these domains? What is the connecting link between the circle form that produced the notion of π and other forms such as rainbows (Adam 2004)?

The stream of digits of π challenges us to try and find a pattern within them. What do they represent? Pi is an irrational number because the digits after the period (of 3.14159...) continue indefinitely, without repeating, and the number cannot be converted into a fraction. The greatest challenge to date, however, has been the race to simply to compute π farther than ever before. The further it has been computed, the more old theories about patterns within π are dispelled and new ones are created. So far, π has been computed to over 50 billion digits. What is the attraction to this number? Is it perhaps the fact that a circle is probably the most perfect and simple form known to human beings? And why does π appear in statistical formulas, biological domains, and in many other spheres of reality? It simply keeps cropping up, reminding us that it is there, and defying us to understand why. Very much like the universe itself, the more technologically advanced we become and as our picture of π grows larger, the more its mysteries grow. There is a beauty to π that keeps our interest in it.

The idea that signs are both simulations (and thus evaluations) of human experience and subsequent locators of new experiences is an extremely problematic one for many philosophers and mathematicians. It begs one of the oldest questions in philosophy and mathematics: Is mathematics invented or is it discovered? As modern physics attempts to develop a "theory of everything" from increasingly abstract mathematics, as financial markets succumb to the mystery of mathematical measures of risk, and as biology employs combinatorial algorithms to unlock the genetic code, it does not seem far-fetched to imagine that numbers do indeed hold the key to the universe. Does the cosmos make mathematics, or does mathematics make the cosmos? The differences in numeration (Roman, decimal, etc.) are, of course, cultured-based and invented, but the similarity of function of all such systems goes beyond culture. All second-order and upper-order systems of representation are really nothing more than substitutes for perception and experience. The human mind creates numbers in the same way that it creates color concepts. Yet the colors, like the numbers, we perceive correspond to something real outside the mind. In this sense, we are discovering numbers all the time.

When all is said and done, the question of where invention ends and discovery begins and vice versa seems to defy a satisfactory answer. The case of the Fibonacci Rabbit puzzle is a truly remarkable one in this regard, because it is, at one level, a simple puzzle, and yet it contains within its solution so many discoveries that it truly boggles the mind to come up with a rational explanation as to why this is so.

The puzzle is found in Fibonacci's *Liber Abaci*, published in 1202. Fibonacci designed his book as a practical introduction to the Hindu-Arabic number system, which he had learned to use during his extensive travels in the Middle East. His method of exposition was based on the creation of puzzles that illustrated how easily the system could be used to solve what would otherwise constitute intractable problems with the Roman numeral system. The puzzle is found in the third section of the *Liber Abaci:*

> A certain man put a pair of rabbits, male and female, in a very large cage. How many pairs of rabbits can be produced in that cage in a year if every

month each pair produces a new pair which, from the second month of its existence on, also is productive?

Going through the solution in class with the students will reproduce Fibonacci's case and argumentation very effectively. There is 1 pair of rabbits in the cage at the start. At the end of the first month, there is still only 1 pair, for the puzzle states that a pair is productive only "from the second month of its existence on." It is during the second month, therefore, that the original pair will produce its first offspring pair. Thus, at the end of the second month, a total of 2 pairs, the original one and its first offspring pair, are in the cage. Now, during the third month, only the original pair generates another new pair. The first offspring pair must wait a month before it becomes productive. So, at the end of the third month, there are 3 pairs in total in the cage—the initial pair, and the two offspring pairs that the original pair has thus far produced. If we keep tabs on the situation month by month, we can show the sequence of pairs that the cage successively contains as follows: 1, 1, 2, 3. The first digit represents the number of pairs in the cage at the start; the second, the number after one month; the third, the number after two months; and the fourth, the number after three months.

During the fourth month, the original pair produces yet another pair. At that point in time the first offspring pair produces its own offspring pair. The second pair produced by the original rabbits has not started producing yet. Therefore, during that month, a total of 2 newborn pairs of rabbits are added to the cage. Altogether, at the end of the month there are the previous 3 pairs plus the 2 newborn ones, making a total of 5 pairs in the cage. This number can now be added to our sequence: 1, 1, 2, 3, 5. During the fifth month, the original pair produces yet another newborn pair; the first offspring pair (now fully productive) produces another pair of its own as well (its second); and now the second offspring pair produces its first pair. The other rabbit pairs in the cage have not started producing offspring yet. So, at the end of the fifth month, 3 newborn pairs have been added to the 5 pairs that were previously in the cage, making the total number of pairs in it: 5 + 3 = 8. We can now add this number to our sequence: 1, 1, 2, 3, 5, 8. Continuing to reason in this way, it can be shown that after twelve months, there are 233 pairs in the cage.

Now, the intriguing thing about this puzzle is the sequence of pairs itself, on a month-by-month basis:

1, 1, 2, 3, 5, 8, 13, 21, 34, 55, 89, 144, 233

Following the reasoning described above, which involves adding pairs in doublets, it becomes obvious that the salient characteristic of this sequence is that each number in it is the sum of the previous two: for example,

2 (the third number) = 1 + 1 (the sum of the previous two); 3 (the fourth number) = 1 + 2 (the sum of the previous two); etc. This pattern can of course be generalized by using simple algebraic notation. If we let F_n represent any "Fibonacci number," and F_{n-1} the number just before it, and F_{n-2} the number just before that, the pattern inherent in the sequence can be shown generally as:

$$F_n = F_{n-1} + F_{n-2}$$

This formula symbolizes the structure hidden in the sequence. It brings out, pedagogically, what algebraic competence is all about. It also shows something that may not have been apparent in the original puzzle, namely that the sequence can be extended ad infinitum, by applying the rule of continually adding the two previous numbers to generate the next:

{1, 1, 2, 3, 5, 8, 13, 21, 34, 55, 89, 144, 233, 377, 610, 987, ...}

Little did Fibonacci know how significant his sequence would become. Over the years, the properties of the Fibonacci numbers have been extensively studied, resulting in a considerable literature. The basic pattern hidden in the sequence was first detected and studied by the French-born mathematician Albert Girard (c. 1595-c. 1632) in 1632. Let's take a rapid look at a few of the discoveries made since then, just for the sake of argument and illustration. In 1753, the Scottish mathematician Robert Simson (1687–1768) noted that, as the numbers increased in magnitude, the ratio between succeeding numbers approached the *golden ratio*, whose value is .618.... The golden ratio results from dividing a line segment in such a way that the ratio of the whole segment to the larger part is equal to the ratio of the larger part to the smaller part. For example, if we take the stretch of numbers in the Fibonacci Sequence starting with 5 and ending with 34, and take successive ratios the results are:

3/5	= .6
5/8	= .625
8/13	= .615
13/21	= .619
21/34	= .618

The golden ratio itself has been found to have astounding reifications. It tends to produce a powerful aesthetic effect—a rectangle the sides of which are in the ratio, tends to be perceived as being the most pleasing of all rectangles, though no one really knows why. Many golden ratios appear in famous paintings, sculptures, and architectural creations. Buildings that incorporate the ratio in their design include the Parthenon (Athens, 400's BCE) and buildings designed in the 1900s by the French architect Le Corbusier. The question this begs is an obvious one: Why would there be a connection between a sequence of numbers produced by a puzzle about copulating rabbits and one of the most enigmatic ratios in the history of human civilization? The plot thickens, so to speak. In the nineteenth century the term Fibonacci sequence was coined by the French mathematician Edouard Lucas, as mentioned, and mathematicians from many domains of inquiry began to discover myriads of patterns hidden within in it. Not only, but stretches of the sequence started cropping up in Nature—in the spirals of sunflower heads, in pine cones, in the regular descent (genealogy) of the male bee, in the logarithmic (equiangular) spiral in snail shells, in the arrangement of leaf buds on a stem, in animal horns, in the botanical phenomenon known as phyllotaxis, in the arrangement of the whorls on a pinecone or pineapple, in the petals on a sunflower, in the branches of some stems, and so on and so forth. In most flowers, for example, the number of petals is one of: 3, 5, 8, 13, 21, 34, 55, or 89 (lilies have 3 petals, buttercups 5, delphiniums 8, marigolds 13, asters 21, daisies 34 or 55 or 89). In sunflowers, the little florets that become seeds in the head of the sunflower are arranged in two sets of spirals: one winding in a clockwise direction, the other counterclockwise. The number in the clockwise is often 21, 34 and counterclockwise 34, 55, sometimes 55 and 89, and sometimes 89 and 144 in the spirals of sunflower heads (Posamentier and Lehmann 2007).

The list of reifications is truly astounding—so much so that a journal, called *The Fibonacci Quarterly*, was established in 1963 to publish new discoveries. Why would the solution to a simple puzzle produce numbers that are interconnected with patterns in Nature and human life? There is, to the best of my knowledge, no definitive answer to this question. Maybe representation is, as mentioned, a discovery tool placed in us by Nature that allows us to come to grips with reality on our own terms. At the same time, it uncovers things about reality that would otherwise remain hidden. As mathematician Ian Stewart (2001: v) has elegantly put it, "simple puzzles could open up the hidden depths of the universe." As a model of artificial rabbit copulation the Fibonacci puzzle has led to an incredible discovery—namely that a simple recursive pattern constitutes the fabric of a large slice of Nature.

Modeling leads to knowledge not because it is designed by instinct to be "knowledge-productive," but because it a semiotic strategy that we use to extract reality from its contexts of occurrence and give it a human form, so that it can be studied on our own terms. The exploratory power of scientific representation, from geometry to quantum physics, suggests that we are probably programmed to discover things through our sign structures, as mainstream structuralists have always maintained. In observing and representing the facts of existence, thus giving them human form, we constantly stumble across hidden patterns in those facts. The Fibonacci Rabbit puzzle episode brings this out perfectly. In the original tale, from which the concept of serendipity is derived, the three princes made their inferences by noticing anomalies that suggested explanations. These spurred their insights. Maybe Fibonacci saw something in a rabbit pen that tickled his fancy and spurred his insight, leading to his puzzle, and to the hidden reifications that it contains. Whatever the truth, Fibonacci's Rabbit puzzle continues to reverberate with implications for the study of the relation between representation and reality. And it makes for a truly interesting anecdote in the classroom, as I have often found out.

4
STORY PROBLEMS

Stand firm in your refusal to remain conscious during algebra. In real life, I assure you, there is no such thing as algebra.

Fran Lebowitz (b. 1951)

Story problems in mathematics, also called word problems, seem constantly to pose learning difficulties for all kinds of students. Recent work on the relation between metaphorical language and PS (for example, Lakoff and Nuñez 2000) suggests that the difficulties stem from an inability on the part of students to decipher the semantic properties of the language in which such problems are cast. In a 2003 pilot study (Danesi 2003), I also found that the metaphorical properties of the language of problems are a constant source of learning difficulty, but that it can be overcome with appropriate semiotic pedagogy. This general line of research will, in my view, shed considerable light on what math competence is and on the crucial role that sign-based processes (including linguistic ones) play in successful PS.

The study of metaphor has a long history behind it in both philosophy and semiotics. But it has been only since the early 1980s that its importance to cognition has been noticed by linguistics. The result has been a new theory of meaning commonly called *Conceptual Metaphor Theory* (CMT). It is this theory that, in my view, has specific implications for imparting skill at solving story problems. CMT is fundamentally a semiotic theory, since it identifies metaphor as a sign-making process that is central to the creation of texts, such as story problems.

The concept that language plays a role in mathematics learning, whether explicitly or implicitly, is of course not new (see, for instance, Klemme 1981, Eccles and Jacobs 1986, Danesi 1987, Varelas 1989, Pimm 1995, MacNamara 1996, Radford and Grenier 1996, Antenos-Conforti, Barbeau, and Danesi 1997, English 1997, MacGregor and Price 1999, Kehle 1999, and Anderson, Sáenz-Ludlow, and Cifarelli 2000). But not till Lakoff and Nuñez's 2000 groundbreaking work did the implications of studying the relation between math and metaphor surface in both mathematics and linguistics. Discussing these is one of the objectives of this chapter, in which I will deal generally with the teaching of story problems from the semiotic perspective.

Defining a Story Problem

A story problem is one that requires solvers to translate the language used by the problem into appropriate mathematical language. As we saw in chapter 2, the translation can be purely arithmetical or, as we saw in chapter 3, it can be algebraic. It can also be geometric, statistical, and so on. In this chapter, only algebraic and geometrical story problems will be discussed.

Consider the following two story problems, found typically in junior and high school textbooks (from Crawford, Dean, and Jackson 1954 and Russell 1960):

> [PROBLEM 4–1]
> Jack is twice as old as John. In four years the sum of the ages of Jack and John will be 20. How old is each at the present time?

> [PROBLEM 4–2]
> The auditorium of a theatre contains 768 seats. If the number of seats in each row is 8 more than the number of rows, find how the seating is arranged.

To solve [4–1] we start by letting John's present age be represented by the variable, x, and Jack's by 2x (twice John's age). In four years, their ages are, thus, x + 4 and 2x + 4. The sum of these ages is 20, which translates into the following equation:

$x + 4 + 2x + 4 = 20$

$3x + 8 = 20$ ← *Simplifying the left side*

Solving the equation by normal algebraic procedures, we discover that John is 4 and Jack is 8. It is always pedagogically useful to go through a solution in its entirety for reinforcement purposes:

$3x + 8 = 20$

$3x = 20 - 8$ ← *Transposing 8, changing its sign*

$3x = 12$ ← *Simplifying the right side*

$x = 4$ ← *Dividing both sides by 3 (= John's age)*

$2x = 2 \times 4 = 8$ ← *Jack's age*

In [4–2], the number of rows can be represented by x. Since there are 8 more seats than rows, the number of seats is x + 8. The statement that the auditorium contains 768 seats may then be translated into the following quadratic equation, since the number of rows × number of seats per row = total number of seats in the auditorium:

$x(x+8) = 768$

$x^2 + 8x = 768$ ← *Carrying out appropriate operations*

$x^2 + 8x - 768 = 0$ ← *Transposing 768, changing its sign*

The last step was designed to get the equation to conform to the standard quadratic form. Using any of the solution techniques for quadratic equations (chapter 3), such as the factoring one, we get:

$x^2 + 8x - 768 = 0$ ← *Transposing 768, changing its sign*

$(x+32)(x-24) = 0$ ← *Factoring*

$x = -32, +24$

We can discard the minus solution. Thus, x = 24. This is the number of rows. The number of seats is 8 more than that, or 32. Therefore, the seating arrangement is 24 rows containing 32 seats each.

These two are examples of closed story problems—story problems involving a straightforward translation of the language of the problem into the language of algebra (as discussed in the previous chapter). However, not all problems are constructed in this clear-cut way. Consider the following two:

> [PROBLEM 4–3]
> A bottle and a cork together cost 55 cents. The bottle costs 50 cents more than the cork. How much does the cork cost?

> [PROBLEM 4–4]
> Sarah is twice the age that Alex was when Sarah was his age. When Alex gets to be Sarah's age, their ages will total 63 years. How old are they?

Problem [4–3] is often misinterpreted by learners, because its language, if read without reflection, typically leads many to the erroneous conclusion that the cork costs 5 cents. If such were the case, then the bottle (costing 50 cents more) would cost 55 cents, and the total cost would be 60 cents. But that is not what the problem states. The appropriate interpretation of the problem can be fleshed out of its statement by setting up an equation. Let x be the price of the cork. This means that x + 50 is the price of the bottle (a representamen which stands for "50 cents more than the price of the cork"). The two prices together add up to 55 cents. The relevant equation is, therefore:

$x + x + 50 = 55$	
$2x + 50 = 55$	← *Simplifying the left side*
$2x = 55 - 50$	← *Transposing 50, changing its sign*
$2x = 5$	← *Simplifying the right side*
$x = 5/2 = 2\ 1/2$	← *Dividing by 2*

The cork thus costs 2 1/2 cents. This means that the bottle, being 50 cents more, cost 52 1/2. Together, the prices add up to 55 (2 1/2 + 52 1/2 = 55). The language in which problem [4–4] is cast is structurally complex and thus very difficult to read correctly. Let x represent Sarah's present age and y Alex's present age. To set up one appropriate equation, it is necessary to interpret the statement "Sarah is twice the age that Alex was when Sarah was his age" correctly. This can be done by rephrasing it with a few concrete examples. Sarah is the older sibling. So, how do we represent Alex's age when Sarah was his age?

- Let's say that Sarah is 40 and Alex 25. How many years ago was Sarah that age? It was (40 - 25) = 15 years ago. We tell the student to note that this is also their present difference in age. Now, how can we figure out Alex's age 15 years ago? We do it by subtracting 15 years (which, again, is the current difference in the two ages) from his present age (25): 25 - 15 = 10.

- Let's say that Sarah is 45 and Alex 29. How many years ago was Sarah that age? It was (45 - 29) = 16 years ago. We comment again that this is also their current difference in age. How can we represent Alex's age 16 years ago? Again, we can do it by subtracting 16 years from his present age (29): 29 - 16 = 13.

- So, if Sarah is x years old and Alex y years old, how many years ago was Sarah that age? It was (x - y) years ago. How can we represent Alex's age (x - y) years ago? As in the concrete examples above, we do it by subtracting (x - y) years (which is the difference in the two ages) from his present age (y): y - (x - y).

We are told that Sarah's current age (x) is twice that age. We can thus represent the entire statement with: $x = 2 [y - (x - y)]$. That is our first equation:

(1) $x = 2 [y - (x - y)]$

$x = 2 [y - x + y]$ ← *Simplifying the right side*

$x = 2y - 2x + 2y$ ← *Simplifying further*

$x = 4y - 2x$ ← *Simplifying further still*

$3x = 4y$ ← *Transposing and simplifying*

Now, we can proceed to set up our second equation. We are told that "When Alex gets to be Sarah's age, their ages will total 63 years." Again, a few concrete examples will be needed to understand what this statement means in algebraic terms:

- Let's say that Sarah is 40 and Alex 25. In how many years will Alex get to be Sarah's age? It will be in (40 – 25) = 15 years from now. As above, we tell the student to note that this is also their current difference in age. How can we represent Sarah's age in 15 years? We can do it by adding 15 years to her present age (40): 40 + 15 = 55. How can we represent Alex's age in 15 years? We can do it as well by adding 15 years to his present age (25): 25 + 15 = 40.

- Let's say that Sarah is 45 and Alex 29. In how many years will Alex get to be Sarah's age? It will be in (45 – 29) = 16 years from now. Again, we emphasize to the student that this is also their current difference in age. How can we represent Sarah's age in 16 years? We can do it by adding 16 years to her present age (45): 45 + 16 = 61. How can we represent Alex's age in 16 years? We can do it by adding 16 years to his present age (29): 29 + 16 = 45.

- In our problem, Sarah's age is x and Alex's is y. In how many years will Alex get to be Sarah's age? It will be in $(x - y)$ years from now. How can we represent Sarah's age in $(x - y)$ years? We can do it by adding $(x - y)$ years to her present age (x): $x + (x - y)$. How can we represent Alex's age in $(x - y)$ years? We can do it by adding $(x - y)$ years to his present age (y): $y + (x - y)$.

We are told that the sum of their ages will be 63. And this allows us to complete our second equation:

(2) $x + (x - y) + y + (x - y) = 63$
$x + x - y + y + x - y = 63$ ← *Removing parentheses*
$3x - y = 63$ ← *Simplifying*

We can now solve our system of equations (chapter 3):

(1) $3x = 4y$
(2) $3x - y = 63$

Without going into the steps here, the solution is x = 28 and y = 21. So, Sarah is 28 years old and Alex 21 years old. As these examples show, solving open or linguistically-complex story problems entails thinking concretely first and then generalizing algebraically from the concrete examples. It involves going from second-order thinking to third-order representation (as it was called in the previous chapter).

Translating statements into algebraic expressions is a basic semiotic substitution process. For example, if we are told that our brother's present salary, which we do not know but wish to know, is increased by 14%, how would we represent his new salary? We start by representing his present salary with a variable, say s. The increase can now be represented by taking 14% of that s, or 14%s. This can be simplified to .14s. Finally, what is his new salary? It is the result of adding the increase, .14s, to his present salary of s, or s + .14s. Using a schema, this can be shown as follows:

Present Salary	*Increase*	*New Salary (Salary + increase)*
↓	↓	↓
s	.14s	s + .14s

Now, if we are told that his new salary is, in actual value, $72,960, we can easily figure out his original salary by setting up an equation. The left side is the new salary, as just represented, and the right side is the actual figure that this represents:

$1.14s = 72{,}960$

We can now discover his original salary by simply solving for s:

$1.14s = 72{,}960$
$114s = 7{,}296{,}000$ ← *Multiplying both sides by* 100
$s = 64{,}000$ ← *Dividing both sides by* 114

The brother's present salary is, therefore, \$64,000. To conclude the presentation, we ask the student to check if the answer obtained by solving the equation is indeed correct. What is 14% of the original salary? It is: .14 × \$64,000 = \$8,960. This is the increase. If we add this to the salary, we get the new one: \$64,000 + \$8,960 = \$72,960.

A Thematic Typology

Classifying story problems thematically is a useful way of alerting students to the linguistic properties that each type generally displays. The main types are: *digit, average, distance, work, money, mixture, age,* and *measurement.* There are other ways to classify story problems, and there are of course other types of story problems. But these are the ones that are the most common, as a survey of texts from various periods that I conducted for the purposes of this book confirmed (see, for example, Moore 1951, Russell 1960, Schneider 1965, Hunter 1965b, Bakst 1967, Perelman 1979, Cathcart, Pothier, and Vance 1994). It is worth going over one problem from each classificatory rubric in detail, so as to get a sense of what it implies in terms of semiotic pedagogy.

Let's start with an example of a digit problem:

> [PROBLEM 4–5]
> The sum of the units and tens digit of a number is 10. If the number is multiplied by 5 the product is 140. What is the number?

There are two ways to solve this problem. One is to set up a system of equations in two unknowns. As the student would know from numeration study (chapter 1), in a two-digit number, the digit in the tens place must be multiplied ten times to get its value. We remind the student with an appropriate schema. What is the value of 25?

25
↙ ↘
(2 × 10)   5 → *Value* → (2 × 10) + 5 = 25

This structural feature is the key to setting up the appropriate system of equations. We let x stand for the tens digit and y for the units digit:

Tens digit ↓ x

Units digit ↓ y

Now, as exemplified above, the value of each of the two digits can be represented as follows:

Value of tens digit	*Value of units digit*
↓	↓
$(10 \times x)$ or $10x$	$(y \times 1)$ or y

So, the value of our number is $10x + y$. By the way, it can be pointed out that this is the general formula for the value of any two-digit number. This can be concretized in chart form as follows:

Two-digit number	=	*Value of tens digit*	+	*Value of units digit*
		↓		↓
		10x	+	y
		↓		↓
25	=	(10) (2)	+	5
38	=	(10) (3)	+	8
94	=	(10) (9)	+	4
98	=	(10) (9)	+	8
...		...		...

Now, the problem tells us that the units and tens digit add up to 10. So, we can easily convert that statement into our first equation:

(1) $x + y = 10$

The second part of the problem tells us that if we multiply the number by 5 the result is 140. What is the value of the number? It is $10x + y$. Multiplying it by 5 can be shown, of course, as $5\,(10x + y)$. The result is 140. Thus, our second equation is:

(2) $5\,(10x + y) = 140$

$10x + y = 28$ ← *Dividing by 5*

We now write down our two equations under each other:

(1) $x + y = 10$
(2) $10x + y = 28$

All that's left to do is solve for x and y, which leads to the solution that $x = 2$ and $y = 8$. The answer is thus 28.

For this type of problem, several translation difficulties tend to characterize the learning process, as I have myself witnessed in classroom situations. So, it is always useful to review relevant patterns, concepts, and formulas associated with digit structure. First, students can be reminded that consecutive integers are represented by adding 1 each time to the previous integer. So, if we let n represent the first integer in the sequence, then the second one in the sequence is $(n + 1)$, the third one is $(n + 1) + 1$, which is simplified to $(n + 2)$, and so on:

First integer	*Second integer*	*Third integer*	*Fourth integer*	…
↓	↓	↓	↓	
n	(n + 1)	(n + 2)	(n + 3)	…

We can also recall the formula for even and odd numbers, for this too might come in handy for solving digit story problems. By definition an even number is a number that is a multiple of two. This means that any number n when multiplied by 2 will always be even. As the student should know (chapter 1), the appropriate formula is 2n:

Even number	*Values for n*	*Result: Always even*
↓	↓	↓
2n	0	$(2 \times 0) = 0$
2n	1	$(2 \times 1) = 2$
2n	2	$(2 \times 2) = 4$
2n	3	$(2 \times 3) = 6$
2n	4	$(2 \times 4) = 8$
…	…	…

By definition, an odd number is any number that is larger or smaller by one of any even number. This is shown by adding or subtracting 1 from the even number formula: $(2n + 1)$ or $(2n - 1)$. Unless the latter is required for some specific reason, it is normal to use the former one:

Odd number	*Values for* n	*Result: Always odd*
↓	↓	↓
2n + 1	0	$(2 \times 0) + 1 = 1$
2n + 1	1	$(2 \times 1) + 1 = 3$
2n + 1	2	$(2 \times 2) + 1 = 5$
2n + 1	3	$(2 \times 3) + 1 = 7$

$2n + 1$	4	$(2 \times 4) + 1 = 9$
...	...	...

Let's now consider average problems. As the student might already know, the average of a list or set of numbers, such as grades in school, is obtained by adding the numbers in the set and dividing the sum by their total number. For example, the average of 3 and 7 is 5, calculated as follows:

$3 + 7 = 10$ ← *Adding the two numbers*
$10 \div 2 = 5$ ←*Dividing the result by the total*

Similarly, the average of 96, 89, 13, and 2 is 50 calculated as follows:

$(96 + 89 + 13 + 2) = 200$ ← *Adding the numbers*
$200 \div 4 = 50$ ← *Dividing the result by the total*

Now, we can consider a typical problem in this category:

[PROBLEM 4–6]
Sarah got back her grades on three tests, math, statistics, and chemistry. She got 82 on the math test, 74 on the statistics test, and 90 on the chemistry test. Her physics test is tomorrow. In order to be the first of her class she needs to get an 85 average on the four tests. What grade must she get on her physics test tomorrow in order to achieve this average?

We start by letting x be the grade she will need to get on her physics test. Now, what will her average be after she gets back her physics test, we can ask? It will be the sum of all the grades (including the unknown one) divided by the number of tests (four in total):

$(82 + 74 + 90 + x)$ ← *Adding the grades*
$(82 + 74 + 90 + x) \div 4$ ← *Dividing the result by the number of tests*
$(246 + x)/4$ ← *Simplifying*
$(x + 246)/4$ ← *Rearranging*

We are told that this should be 85 (ideally). So, let's equate it to 85 and then, of course, solve for x:

$(x + 246)/4 = 85$
$x + 246 = 340$ ← *Multiplying by 4*
$x = 340 - 246$ ← *Transposing 246, changing its sign*
$x = 94$ ← *Simplifying the right side*

This tells us that Sarah must get 94 on her physics test in order to get an average of 85 overall. Let's check and see if this is correct:

$(82 + 74 + 90 + 94) = 340$ ← *Sum of the grades*
$340 \div 4 = 85$ ← *Average of the four grades*

Now, whether Sarah will get 94 on her physics test or not is another matter. All we were asked to do (we remind the student) is determine what grade she would need to get on the test in order to get an 85 average for the four tests. And we now know what this is.

A third, and rather pesky, type of story problem is known as a *distance* or *uniform motion* problem. The formula *distance* (D) *is equal to the rate* (R) *multiplied by the time* (T) is always involved. The student should be reminded of it:

$$D = R \times T$$

For example, if we travel 50 miles in one hour in a car and take 14 hours, what total distance have we traveled? It is of course: 50 × 14 or 700 miles. This can be shown as follows:

$D = R \times T$
$D = (50)(14)$
$D = 700$

If we are told that a 1000 km trip took 10 and a half hours, what was the rate of speed of the driver (on average of course)? It was 1000 ÷ 10 1/2 hours or 95.24 kilometers per hour. This too can be shown with the formula:

$D = R \times T$
$1000 = (R)(10\ 1/2)$ ← *Plugging in 1000 for D and 10 1/2 for T*
$1000 = (R)(10.5)$ ← *Converting 10 1/2 to decimal form*
$1000 \div 10.5 = R$ ← *Dividing both sides by 10.5*
$R = 95.24$ ← *Simplifying*

And, if we are asked to calculate how much time it would take to get to a place that was 550 miles away driving at an (average) rate of 50 miles an hour, the answer would be, of course, $550 \div 50 = 11$ hours. Using the formula, we show this as follows:

$D = R \times T$

$550 = (50)(T)$ ← *Plugging in 550 for D and 50 for R*

$550 \div 50 = T$ ← *Dividing both sides by 50*

$T = 11$ ← *Simplifying*

Some problems will present factors that affect such things as the rate. One of these is the "wind factor." If an airplane flies with the wind at its back, the result is that the wind speed is added to the airplane's speed. For example, if the plane flies at 400 miles per hour and is propped up by a wind at its back of 50 miles per hour, its overall speed for the journey is 400 + 50 = 450 miles per hour. However, if an airplane flies against the wind, the result is that the wind speed is subtracted from the airplane's speed. Again, if the plane flies at 400 miles per hour and the wind blowing against it is 50 miles per hour, its overall speed for the journey is 400 - 50 = 350 miles per hour.

At this point we can introduce a typical distance problem:

[PROBLEM 4–7]

In a famous driving competition, Jill and David drove the same course at (average) speeds of 51 miles per hour and 54 miles per hour respectively. Jill got to the end of the course in 40 minutes. How many minutes did it take David to get there?

Since we do not know how long it took David to get to the end of the course, we let his time be x minutes. Now, we have to figure out how long the course is. We are told that Jill did it in 40 minutes driving at 51 miles per hour. Incidentally, 40 minutes is 40/60 of an hour. So:

$D = R \times T$

$D = (51)(40/60)$ ← *Plugging in the values*

$D = 34$ miles ← *Simplifying*

Now that we know the distance, we can compute the time taken by David to complete it, since we also know his rate, 54, and that his time is x, which is x/60 of an hour. So:

$D = R \times T$	
$34 = (54)(x/60)$	← *Plugging in the values*
$34 = 54x/60$	← *Carrying out the operations*
$34 \times 60 = 54x$	← *Multiplying both sides by* 60
$2040 = 54x$	← *Carrying out the operations*
$x = 37.77 \approx 37.8$	← *Dividing both sides by* 54

Thus, it took David almost 37.8 minutes to get to the finish line. Another common type of story problem involves work rates. This type is solved by comparing rates on a per hour, per job, etc. basis. The crucial thing for the student to understand is that if someone takes 5 hours to complete a job, working steadily that person can complete 1/5 of the job in one hour and, of course, 2/5 in two hours, 3/5 in three hours, 4/5 in four hours completing it in five hours, when he or she will have completed 5/5 of the job. This is, of course, the entire job. Below is a typical example of a work problem:

[PROBLEM 4–8]
If it takes Alex 3 hours to complete a job of painting a room and it takes Shamila only 2 hours to paint the same room, how long would it take to paint the room if they work on it together (at the same time)?

The solution is based on considering the hourly rate of work of each person. Alex can complete the job in 3 hours. So, on average he completes, 1/3 of the job in one hour (and of course, 2/3 in two hours and 3/3, or the whole job, in three hours). Shamila completes the job in 2 hours. So, on average she completes 1/2 of the job per hour.

We are asked to determine how much time it would take when the two are painting together. Let's represent this time with x. The two of them working together will complete the job in x hours. So, in one hour they would complete 1/x of the job (on average). We know that Alex completes 1/3 and Shamila 1/2 working alone in one hour. Summing their two rates, we get:

$$\frac{1}{3} + \frac{1}{2} = \frac{1}{x}$$

The solution is: x = 6/5. This means that it will take Alex and Shamila 6/5 hours, or 1 hour and 12 minutes, to complete the painting job working together.

Another common type of story problem involves money—wages, costs, prices, investments, taxes, and so on. The denomination used here is the dollar, although the same principles and techniques apply to any monetary denomination. Here is a typical problem:

> [PROBLEM 4–9]
> At a supermarket, apples can be purchased on sale for $0.15 each and pears for $0.20 each. At these bottom bargain rates, Chris purchased a bag of 21 apples and pears together for $3.80. How many pears did he purchase?

This can be solved in two ways. Let's use two unknowns, letting the number of apples purchased by Chris be a and the number of pears p. Now, how much of the $3.80 was for the apples? Well, since we know that one apple costs $0.15 cents, then a of them would cost in total a × .15 or .15a. How much was for the pears? One pear costs $0.20 cents, so p of them would cost in total a × .20 or .20p. Now we can represent the total cost of the purchase concretely as follows:

Cost of the apples	+	*Cost of the pears*	=	*Total cost*
↓		↓		↓
.15a	+	.20p	=	3.80

This is our first equation. We are told that there are 21 items of fruit in the bag. Since Chris bought a apples and p pears, then these numbers add up to the total: a + p = 21, which is our second equation:

(1) .15a + .20p = 3.80
(2) a + p = 21

Solving the equations, we get a = 8, p =13, which means that there were 8 apples and 13 pears in the bag. We could also solve this problem by letting x stand for the number of pears and (21 - x) for the number of apples (since that would be the number of apples in the bag of 21 when the number of pears is taken away). To show students how to set up the equation, we use the same type of concretization technique:

Cost of the apples	+	*Cost of the pears*	=	*Total cost*
↓		↓		↓
.15 (21 - x)	+	.20x	=	3.80

Our equation would thus be:

$.15\,(21 - x) + .20x = 3.80$

Now, here is a problem involving investment:

> [PROBLEM 4–10]
> Alex invests $500 more at 5% than he already has invested at $4\frac{1}{2}$%. The total annual income (yield) from the two investments is $329. How much has Alex invested at each rate?

The amount Alex invests at $4\frac{1}{2}$% can be represented by x. Since he invested $500 more at 5%, the appropriate expression is (x + 500). What is the annual income from the investment of x? It is, of course, $4\frac{1}{2}$% of x, or, in simplified form, .045x. The annual income from the investment of (x + 500) is 5% of (x + 500), or, in simplified form, .05 (x + 500). The income from these two investments is $329. Using concretization, this can be shown as follows:

Investment on x	+	*Investment on* (x + 500)	=	*Annual income*
↓	+	↓	=	↓
.045x	+	.05 (x + 500)	=	329

The appropriate equation is, thus:

$.045x + .05\,(x + 500) = 329$

Solving for x, we find that Alex invested $3,200 at $4\frac{1}{2}$% and $3,700 at 5%. Another common category of story problem involves mixture, percentages, solutions, and the like. Here's a typical example of this kind of problem:

> [PROBLEM 4–11]
> A mixture of 12 ounces of vinegar and oil consists of 40% vinegar (by weight). How many ounces of oil must be added to that mixture to produce a new mixture with only 25% vinegar in it?

The number of ounces of oil to be added can be represented by x. The total number of ounces in the new mixture will thus be (12 + x). Now, it is easy to figure out how many ounces of vinegar there are in the original mixture, since we are told that it is 40% of the 12 ounces:

$$(.40)\,(12) = 4.8 \text{ ounces of vinegar}$$

How many ounces of vinegar are there going to be in the new mixture? The new mixture will have (12 + x) ounces in it, and 25% will consist of vinegar. This can be represented as follows:

$$.25\,(12 + x)$$

This does not alter the actual amount of vinegar in the mixture, since we have added oil, not vinegar to it. So we know that it must equal the original 4.8 ounces. The only thing that will change is its percentage (since more oil has been added to the mixture). So:

$$.25\,(12 + x) = 4.8$$

Solving, we get $x = 7.2$. This means that 7.2 ounces of oil must be added to produce the new percentage of vinegar in the mixture.

Age problems constitute a very common category of story problems. There are two common sense things for the student to keep in mind when setting up equations in this case. If someone's age now is x then twenty years ago it was (x - 20). If students do not see this, we can ask them to consider concrete examples, as was done above for problem [4–4]:

- If someone is 52, how old was that person twenty years ago? He or she was, of course, $52 - 20 = 32$.

- If someone is 34, how old was that person twenty years ago? He or she was $34 - 20 = 14$.

- In general, if someone is x years old, that person was x - 20 years old twenty years ago. Even more generally, if someone is x years old, then that person was x - y years old y years ago. Thus, if asked to calculate or represent what someone's age was a number of years ago, we subtract the number from his or her present age.

Another thing to be kept in mind is that if someone's age now is x, then twenty years from now it will be (x + 20). If students do not see this, we can use the same concretization technique:

- If someone is 32, how old will that person be twenty years from now? He or she will be 32 + 20 = 52.

- If someone is 14, how old will that person be twenty years from now? He or she will be 14 + 20 = 34.

- In general, if someone is x years old, that person will be x + 20 years old twenty years from now. Even more generally, if someone is x years old, then that person will be x + y years old y years from now. Thus, if asked to calculate or represent what someone's age will be a number of years from the present, we add the number to his or her present age.

Here is an example of an age problem:

> [PROBLEM 4–12]
> The ages of John and Mary together are 26. If we double their ages, John's age (who is older) will exceed Mary's by 8 years. How old is each one?

Let x be John's age, then (26 - x) is Mary's age, as students should be able to easily see by now. So we double both their ages, which can be represented concretely as follows:

John's age doubled	*Mary's age doubled*
↓	↓
2x	2 (26 - x) = 52 - 2x

We are told that John's doubled age will exceed (be greater than) Mary's doubled age by 8. In other words, the difference between the two will be 8. So, the appropriate equation is:

$$2x - [(52 - 2x)] = 8$$

Solving for x, we find that John is 15 and Mary is 11. Finally, measurement story problems involve some basic knowledge of geometry, especially facts and theorems such as the following, which the teacher should ensure are known by the students beforehand:

- The area (A) of a rectangle is the product of the length (l) times the width (w): $A = l \times w$
- The area of a square is also the product of the length times the width, but since the two are equal by definition, the area is $A = x \times x = x^2$, where x is the length of a side of the square.
- The area of a triangle is $\frac{1}{2} \times b \times h = \frac{1}{2}bh$, where b is its base and h its height (perpendicular).

Here is an example of a measurement problem:

> [PROBLEM 4–13]
> A rectangle is 8 feet longer than its width. Its area is 308 square feet. What are the dimensions (length and width) of the rectangle?

In all measurement problems, it is essential to draw a diagram that represents the problem, if no diagram is provided. Let's draw our rectangle on which we let the width be represented by x and the length by x + 8, since we are told that it is 8 feet longer than the width:

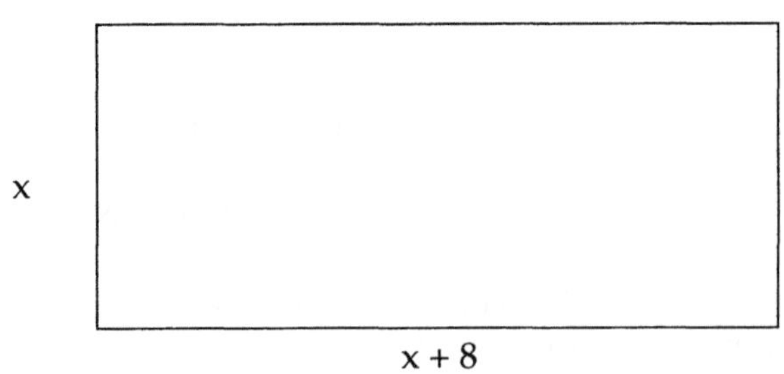

The area is, therefore: $x(x + 8) = x^2 + 8x$. We are told that this is equal to 308 square feet. So, the appropriate quadratic equation is:

$$x^2 + 8x = 308$$

$x^2 + 8x - 308 = 0$ ← *Transposing 308, changing its sign*

$(x + 22)(x - 14) = 0$ ← *Factoring*

$x = -22, +14$

We can discard -22 as a solution, since the width must be a positive number. So, the width is 14 feet and the length is 22 feet.

Inter-Code Translation

Among the first to describe solutions to story problems as an exercise in translation from regular language to algebraic language, that is, from one from code to another code, was none other than that great English mathematician and physicist, Sir Isaac Newton (1642–1727)) in his *Artihmetica Universalis* (Perelman 1979). The following is a model problem used by Newton himself to exemplify this very point:

In ordinary language	*In the language of algebra*
A merchant has a certain sum of money.	x
During the first year he spent 100 pounds.	(x - 100)
To the remaining sum, he then added one third of it.	(x - 100) + (x - 100)/3 = (4x - 400)/3
During the next year he again spent 100 pounds.	(4x - 400)/3 - 100 = (4x - 700)/3
And increased the remaining sum by one third of it.	(4x - 700)/3 + (4x - 700)/9 = (16x - 2800)/9
During the third year he again spent 100 pounds.	(16x - 2800)/9 - 100 = (16x - 3700)/9
After he added to the remainder one third of it,	(16x - 3700)/9 + (16x - 3700)/27 = (64x - 14800)/27
his capital was twice the original amount.	(64x - 14800)/27 = 2x

To determine the original capital, we have only to solve the last equation. It is obvious, as Newton demonstrated, that the process of setting-up equations reduces to an *inter-code* system, whereby ordinary language is converted to algebraic language by a series of substitutive operations. The task, then, is to determine what those operations are. By observing the problem-solving process closely, it is possible to posit two general inter-code operations: *analysis* and *transference* (Antenos, Barbeau, and Danesi 1998).

Analysis is an intratextual operation; that is, it is an operation performed on the target or source language (SL) text. It can be characterized as an essentially interpretive process, involving an understanding of the SL concepts as expressed in terms of the SL code. Analysis constitutes the obvious first step in setting up equations, since the problem-solver must, above all else, understand *what* is required in terms of *how* it is expressed. The following five problems will suffice to illustrate some of the different modalities involved in the analysis of the SL text (taken from Moore 1951, Mott-Smith 1954, Russell 1960, Bakst 1967):

> [PROBLEM 4–14]
> Alice is three times as old as Joan. In eight years, the sum of their ages will be 28. What are their present ages?

> [PROBLEM 4–15]
> Young Mrs. Martin called her mother on the telephone to ask help in calculating her husband's income. "This morning he told me that since we were married (and that was three years ago) we spent four-tenths of his yearly income on rent and the upkeep of the house, one-fourth on clothes and food, and one-tenth on other incidentals. He never told me what his income was. But only a few days ago I found his bankbook, and he had saved $1,500. Don't you think that I ought to know what my own husband earns?" Can you find out?

> [PROBLEM 4–16]
> If a brick balances evenly with three-quarters of a pound and three-quarters a brick, what is the weight of the whole brick?

> [PROBLEM 4–17]
> In how many minutes after 8 o'clock will the minute overtake the hour hand?

The conceptual content in [4–14] can be analyzed with a simple tree diagram that shows how the parts connect to each other. This is what was done in the problems discussed previously above:

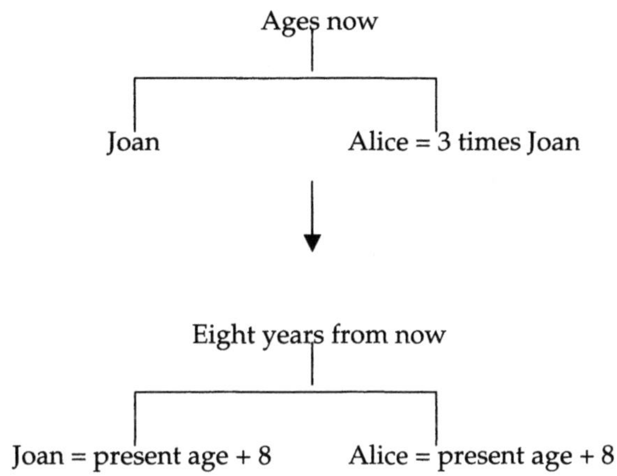

The analysis of [4–15] can be represented with the following tree diagram:

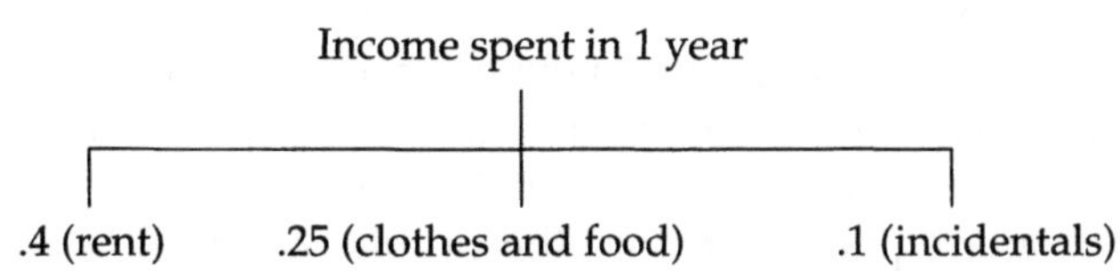

The catch here is that the amount saved refers to a three-year period, whereas the statement concerning expenditures refers to a one-year period.

In [4–16], which is a version of problem [1–4] in the first chapter, the compound statement "three-quarters of a pound and three-quarters of a brick" may lead, if improperly analyzed, to the erroneous conclusion that the brick weight is 3/4 + 3/4 pounds. A correct analysis can be shown with the following tree diagram:

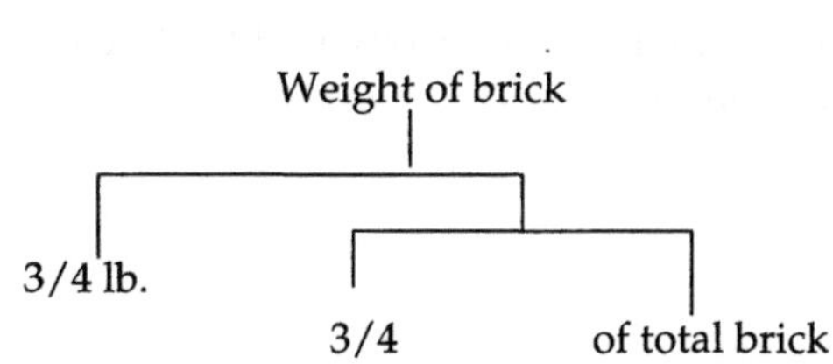

The right-hand branch shows the key to the solution of this problem, namely that 3/4 *of* some weight is involved. The analysis of problem [4–17] can be represented diagrammatically as follows, based on observing how the hands of a clock move with respect to each other:

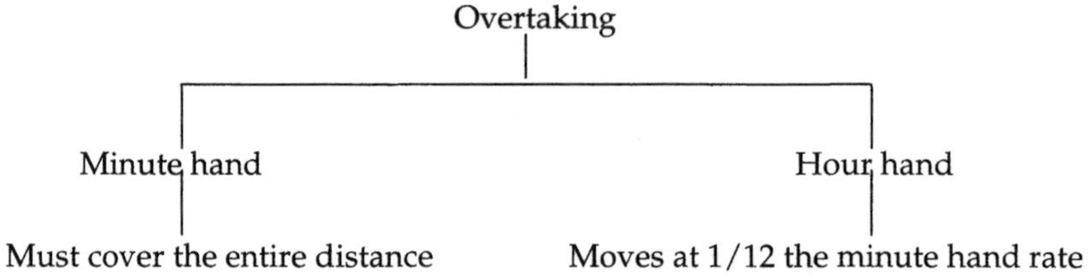

Generally speaking, in straightforward inter-code translations the substitutive process is one-to-one: for example, in translating Roman numerals into decimal ones, the translator determines what numerical value is involved in the source text and then what the appropriate signifiers (digits) of the target code are to be used to represent it in the target text. But in other kinds of translation, the procedure is much trickier. To quote Kasner and Newman (1940: 158): "It is common experience that often the most formidable algebraic equations are easier to solve than problems formulated in words. Such problems must first be translated into symbols, and the symbols placed into proper equations before the problems can be solved." Once the content of the verbal problem has been properly analyzed and appropriately represented, the task of the problem-solver consists in giving it an algebraic form. This is a transference operation, by which the diagram's components are converted into algebraic symbols and expressions. As Schneider (1965: vi) suggests, a simple conversion system can be normally constructed on the basis of either an overtly-stated, or implied mathematical concept. Thus, for example, the operation of addition is usually implied by words such as *and, sum, total, add together*, etc., in the SL text and is translated, of course, by the sign +; similarly, subtraction is normally implied by expressions such as *less, from, take away, difference, is greater than, subtract*, etc., and is translated by -; and so on.

The diagrammatic analysis of problem [4–14] can now be converted into algebraic form as follows:

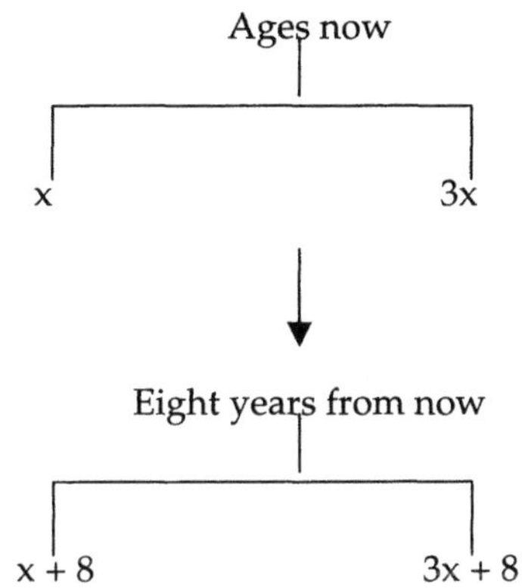

The problem states that in eight years their ages will total 28, which, of course, implies an addition symbol. This leads to our equation and the solution:

$x + 8 + 3x + 8 = 28$

$4x + 16 = 28$ ← *Grouping*

$4x = 28 - 16$ ← *Transposing 16, changing its sign*

$4x = 12$ ← *Simplifying*

$x = 3$ ← *Dividing by 4*

This means that Joan is 3 and Alice, being three times as old, is 9. The diagram of [4–15] can be converted to algebraic form as follows, letting the total income of the husband be x:

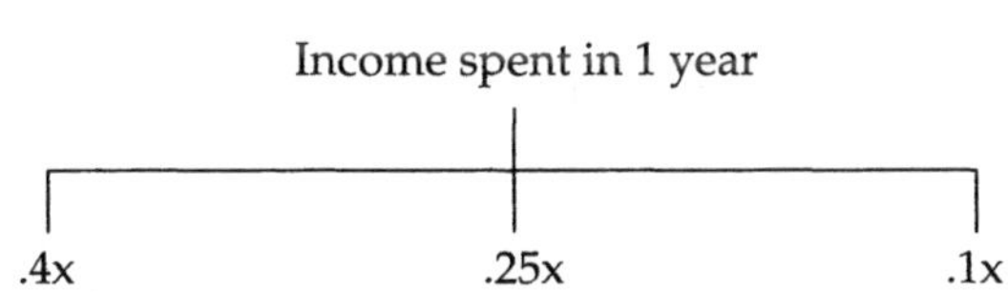

The total income spent in one year is thus: $.4x + .25x + .1x$. We are told that the husband saved $1,500 in three years. So, in one year he saved $500. Thus, his income in one year would equal expenses + savings:

$x = (.4x + .25x + .1x) + 500$

↑ income ↑ expenses ↑ savings

x = .75x+ 500	← *Grouping*
x - .75x = 500	← *Transposing .75x, changing its sign*
.25x = 500	← *Simplifying*
x = 2,000	← *Dividing by .25*

Thus, the husband earned $2,000 each year, spending $1,500 and saving $500 per year. The diagrammatic analysis of [4–16] can be converted to its corresponding algebraic form as follows:

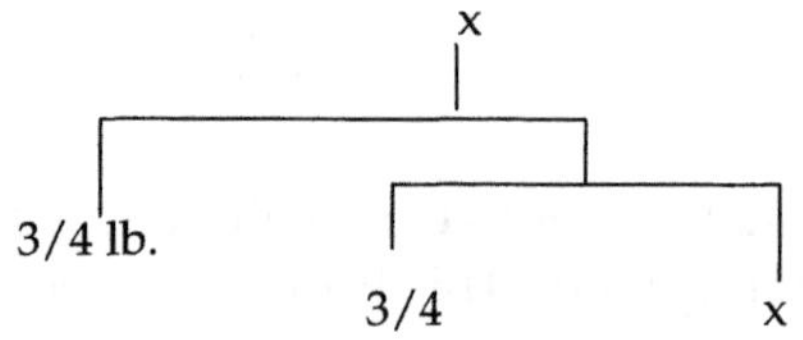

Thus, the total weight of the brick (x) can be expressed as:

x = 3/4 + 3/4x	
4x = 3 + 3x	← *Multiplying by 4*
4x - 3x = 3	← *Transposing 3x, changing its sign*
x = 3	← *Simplifying*

This means that the brick weighed 3 lb. The diagrammatic analysis of [4–17] can be converted to algebraic language as follows (shortening the diagram somewhat):

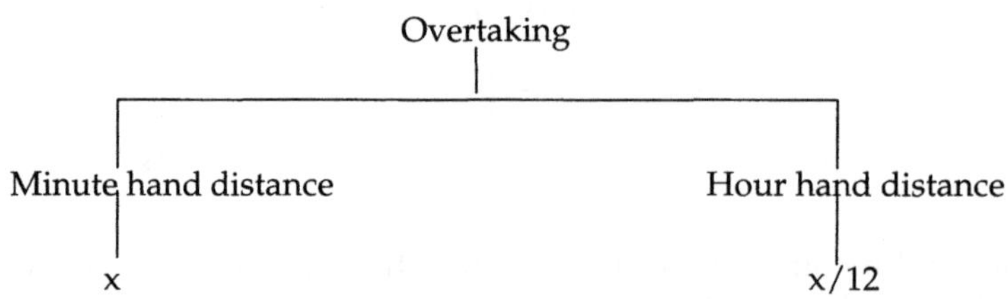

The x/12 can be explained as follows. Since the hour hand covers five units in one hour, in going from one hour to the next, the minute hands covers 60 in the same time period. This is 1/12 the rate of the minute hand—hence the x/12. Now, there are 40 minute divisions in difference between the hour and minute hands at 8:00 (as students can confirm for themselves). Thus, the difference in distance between the two hands can be expressed as:

$x - x/12 = 40$

$12x - x = 480$ ← *Multiplying by 12*

$11x = 480$ ← *Simplifying*

$x = 43.6$ ← *Dividing by 11*

This means that the minute hand will overtake the hour hand in 43.6 minutes. In sum, transference consists in converting SL concepts as expressed in specific kinds of words into algebraic signs.

Conceptual Metaphor Theory

As a result of using story problems in experimental courses at the University of Toronto (mentioned in the preface of this book), I decided to carry out a pilot study for teaching mathematics in middle school that addressed the problem of metaphorical meaning in the language of such problems. The study produced highly interesting results (Danesi, 2003). So, in this final section of the present chapter my aim is to discuss the implications of that study in a more general light—implications supported by a follow-up project conducted in early 2007 (Danesi 2007a) in which the same pedagogical framework, based on Conceptual Metaphor Theory (CMT), was utilized to teach two classes of self-defined math phobics at the University of Toronto. The idea was to compare the outcomes of the math phobic group to the findings of the original 2003 case study so as to determine if the CMT framework was capable of producing replicable results or not.

As mentioned at the beginning of this chapter, CMT is one of the various semiotically-oriented proposals that have started to infiltrate math education circles over the last few decades as viable alternatives to mainstream forms of pedagogy. CMT is potentially a very powerful teaching tool because it unmasks hidden metaphorical concepts in all kinds of texts, from literary to mathematical and scientific ones. In the area of mathematics, the model posits that many concepts in mathematics, from purely arithmetical ones to highly advanced ones, are structured rhetorically and, thus, are best deciphered and taught by making direct reference to their rhetorical structure (Reed 1994, English 1997, Lakoff and Nuñez 2000).

From the extensive research on the presence and function of metaphor in language and other symbolic systems since at least the mid-1970s, it has become apparent that rhetorical thinking constitutes a fundamental cognitive substratum on which many of our common abstractions are built, including mathematical ones. The pedagogical implications of this general finding can no longer be ignored. The results of the studies in this area suggest, in fact, that a pedagogical framework incorporating CMT is not only useful in helping students with difficulties in math problem-solving, but probably generalizable to all types of learners, learning tasks, and educational environments.

Metaphor has always been a central area of interest and research in semiotics. Charles Peirce, one of the modern-day founders of the discipline (chapter 1), saw it, in his various writings, as the basis for inferential reasoning in the sciences and as the cognitive guide in the design of diagrams that are created on purpose to illustrate scientific theories. Rarely, however, has the role of metaphor in learning been examined from an educational angle, probably because it did not catch the attention of educators until after the publication of Lakoff and Johnson's highly popular and widely influential book *Metaphors We Live By* (1980). That work presented in a coherent fashion a theory of conceptualization that has since found its way into various models of language, cognition, cultural symbolism, and learning processes. In the early 1990s, I attempted to incorporate CMT into a model of teaching and learning foreign languages (Danesi, 1994, 1995) that produced a number of positive research findings (see, for example, Kövecses and Szabo 1996, Kecskes 2000, Holme 2004, and Velasco Sacristán 2005). Some disagreement did emerge as to the coverage of the framework (Valeva 1996, Boers 2000). But the point of contention was, in my view, a moot one, since the definition of an abstract concept as having rhetorical structure is not in question within CMT itself; nor is the possibility that some forms of abstract conceptualization may have other (non-rhetorical) psychological sources.

Understanding what a concept is and how it is learned has been the target of much research and debate within psychology, linguistics, semiotics, and many other human sciences since their inception. Concepts fall, by and large, into two basic domains—concrete and abstract. A concrete concept is one whose referent is demonstrable and observable in a direct way, whereas an abstract concept is one whose referent cannot be demonstrated or observed directly. So, for example, the word *car* stands for a concrete concept because its referent, "a self-propelled land vehicle, powered by an internal-combustion engine," can easily be demonstrated or observed in the physical world. The word *love*, on the other hand, represents an abstract concept because, although "love" exists as an emotional phenomenon, it cannot be demonstrated or observed directly—that is, the emotion itself cannot be extracted from the behaviors, states of mind, etc. that it produces.

The distinction between concrete and abstract concepts is, needless to say, a conveniently reductive one. In actual fact, there are many degrees of concreteness and abstraction in conceptualization that are influenced by various kinds of psychological and social factors, as the enormous literature in the semiotic, social, and cognitive sciences has documented over the last century. Suffice it to say here that our brain seems structured to organize most of the raw, unorganized information that comes from perception into useful concepts, most of which can be classified in terms of a concrete versus abstract dichotomy.

The question of how to teach concrete concepts in subjects ranging from foreign languages to mathematics has always had a fairly straightforward answer. It is assumed that they are best taught in conformity with their referential nature—namely, by demonstrating them through explanatory and ancillary devices that allow their referential structure to be shown in some concrete way. By and large, this type of pedagogy has always produced, and will continue to produce, good results in all kinds of classrooms. It has been called concretization in this book. Essentially, it is the technique of demonstrating a referent in concrete terms. In mathematics, as we have seen, this generally involves the use of visual representamina (diagrams, charts, etc.) that allow the teacher to demonstrate a concrete concept in a direct way. For example, we have used the number line several times as a representamen to show the difference between positive and negative numbers, to show how addition and subtraction unfold conceptually, and so on.

Visual devices are easy to comprehend because they concretize concepts such as that of number itself by showing a distance between any two numbers as being constant. Since students are also familiar with such devices as thermometers, the number line can also easily be associated with real-life exemplars to impart the concept of negative number. This technique has been called contextualization in this book. The thermometer is really a number line—a line on which all the numbers, positive and negative, can be displayed with 0 in the middle as the boundary point between them. The negative numbers are placed to the left and the positive ones to the right of 0. Zero is not defined as either negative or positive. This makes it the natural mid point on many scales, such as thermometers. In this way, students can be shown concretely that negative numbers have real-life functions—any temperature marking below or to the left of the zero point on a thermometer is indicated by a negative number; any number marking above or to the right of the zero point is indicated instead as a positive number.

The dilemma in education, it would seem, has always lied specifically in the teaching of abstract concepts (not concrete ones). It is precisely in attempting to resolve this dilemma that CMT presents itself as a potentially useful framework because it suggests that many abstract concepts can be concretized in exactly the same way as concrete ones such as positive and negative numbers. Why is this so? The answer is suggested by the research findings within CMT itself which show, essentially, that many common abstract concepts are conceptualized in terms of concrete ones, revealing their concrete referential structure through the medium of metaphor (Pollio, Barlow, Fine, and Pollio 1977, Langacker 1987, 1990, 1999, Gibbs 1994, Dirven and Verspoor 1998, Lakoff and Johnson 1999, Leezenberg 2001, Fauconnier and Turner 2002, Kövecses 2002, Croft and Cruse 2004, Geeraerts 2006,

Knowles and Moon 2006). It is for this reason that Lakoff and Johnson (1980) decided to rename abstract concepts *conceptual metaphors*, defining them as generalized metaphorical formulas. For example, the metaphor "John is a gorilla" is really a token of something more general, namely, *people are animals*. This is why we can also say that *John* or *Mary* or whoever is a *gorilla, snake, pig, puppy,* and so on. Each specific linguistic metaphor ("John is a gorilla," "Mary is a snake," etc.) is not an isolated example of poetic fancy. It is really an instantiation of a more general metaphorical idea—*people are animals*. Such formulas are what Lakoff and Johnson call conceptual metaphors.

Each of the two parts of the conceptual metaphor is called a domain: *people* is called the target domain because it is the abstract topic itself (the "target" of the conceptual metaphor); and *animals* is called the source domain because it is the class of concrete vehicles that undergird the metaphorized idea (the "source" of the abstract concept). An abstract concept in CMT is defined simply as a "mapping" or an "extension" of one domain onto the other. This theory suggests that abstractions are conceptualized systematically through such extensions and that specific metaphors are traces to the target and source domains. So, when we hear people talking, for instance, of *ideas* in terms of *geometrical figures* and *relations*—"Those ideas are circular;" "I don't see the point of your idea;" "Her ideas are central to the discussion;" "Their ideas are diametrically opposite;" etc.—we can now easily identify the two domains as *ideas* (= target domain) and *geometrical figures/relations* (= source domain) and, therefore, identify the underlying metaphorical formula as: *ideas are geometrical figures and relations.*

We do not detect the presence of conceptual metaphors in discourse because of repeated usage and because they are guided by unconscious processes called *image schemata* (Lakoff 1987, Johnson 1987). These are conceptual oppositions, as semioticians would call them. For example, the image schema *up-versus-down* is the underlying opposition that guides the conceptualization of such abstractions as mood and income: "I'm feeling *up*;" "She's feeling *down* today;" "My income *rose* (went *up*) last year; "Her salary went *down* when she changed job;" etc. Image schemata are so deeply rooted that we are hardly ever aware of their control over conceptualization. For example, it is an unconscious image schema that leads us to conceptualize certain abstractions as memory and anger as entities occurring in containers "I'm *full* of memories;" "I am *over-brimming* with anger;" etc.

Image schemata provide the key to the concretization process in the teaching of story problems, suggesting how to design the form or make-up that a diagram or chart should assume. Consider a simple story problem in algebra that has been used already various times in this book:

[PROBLEM 4–18]

John is five years older than Mary. Four years from now, he will be twice her age. What is the present age of each?

A CMT analysis of this problem shows that it is constructed on the basis of two image schemata of *time*. They are the same ones that undergird the use of the prepositions *since* and *for* in common sentences such as the following:

I have been living here *since* 2000.
I have known Lucy *since* November.
I have not been able to sleep *since* Monday.
I have been living here *for* fifteen years.
I have known Lucy *for* nine months.
I have not been able to sleep *for* five days.

An analysis of the complements that come after *since* reveals that they are metaphorical constructions produced by the image schema of time as a "point" on a "timeline:" *2000, November, Monday*. In contrast, the complements that follow *for* reflect a conception of time as a "quantity:" *fifteen years, nine months, five days*. These two conceptual metaphors—*time is a point* and *time is a quantity*—are implicit in the formulation of the above problem. Arguably, when the conceptual structure of the problem is deciphered, solving it becomes a straightforward process. If we let x represent Mary's present age, then John's present age would be represented by x + 5. This is so, because time is a point on a timeline diagram on which John's age-point is "5 points" to the right of Mary's age-point. Representing John's and Mary's ages "four years from now" entails moving John's and Mary's individual age-points "to the right by four" on the timeline. This translates into the algebraic expression x + 4 for Mary and $x + 5 + 4$ (= $x + 9$) for John:

Mary's age four years from now, shown on a year-by-year basis

Timeline	0	1	2	3	4
	↓	↓	↓	↓	↓
Representation	x	x + 1	x + 2	x + 3	x + 4
	↓	↓	↓	↓	↓
Meaning	age now	age one year from now	age two years from now	age three years from now	age four years from now

John's age four years from now, also shown on a year-by-year basis

Timeline	0	1	2	3	4
	↓	↓	↓	↓	↓
Representation	x + 5	x + 5 + 1 = x + 6	x + 5 + 2 = x + 7	x + 5 + 3 = x + 8	x + 5 + 4 = x + 9
	↓	↓	↓	↓	↓
Meaning	age now	age one year from now	age two years from now	age three years from now	age four years from now

Finally, to set up an algebraic relation between the two ages, it is necessary to "shift source domains," so to speak, to *time is a quantity*. John's age is quantifiable as twice that of Mary's age. This is, metaphorically speaking, a difference in the size of the container schemata that hold the ages (conceptualized as entities). The step is, thus, a straightforward one—expressing in algebraic notation the overall "metaphorical relation." We know that John's age four years from now is (x + 9). We also know that this is twice Mary's age at that time, that is (x + 4) + (x + 4), or 2x + 8. This leads to the following equation:

$x + 9 = 2x + 8$

$x - 2x + 9 = 8$ ← *Transposing 2x, changing its sign*

$-x + 9 = 8$ ← *Simplifying the left side*

$-x = 8 - 9$ ← *Transposing 9, changing its sign*

$-x = -1$ ← *Simplifying the right side*

$x = 1$ ← *Dividing by -1*

Solving the equation reveals that Mary is one year old and John six. Every story problem in English that involves *time* can be represented and explained in an analogous way, that is, in terms of the *timeline* and *quantity* image schemata. This was brought out by an extensive analysis of all age problems in algebra textbooks in CMT terms. In effect, a CMT analysis of the language of the problem provides the necessary insight on how to represent it in concrete terms for pedagogical purposes. The concretization process in this case involves figuring out the source domains used in the formulation of the problem. These involve image schemata such as containers, lines, impediments, substances, etc. As Lakoff and Nuñez (2000) have suggested, success at mathematics entails knowledge of how these are used to encode mathematical concepts. Some mathematicians are finally starting to realize the power of this insight. Keith Devlin (2005), for instance, has recently suggested that metaphor is the source of mathematical concepts because it is the form of reasoning that allows us to understand something new and unfamiliar in terms of something familiar and already understood.

As the foregoing discussion was intended to suggest, identifying the image schemata used in constructing story problems and converting them into teaching devices such as diagrams is the cornerstone principle in a CMT pedagogical framework. Consider, again, the problem discussed above. As the relevant research projects discussed above brought out, the main difficulty experienced by students initially in solving it (and others like it) stemmed from an inability to represent it in a concrete way, that is, in terms of the two image schemata of time implicit in the problem. In CMT terms, the students did not know how to concretize the source domains. Indeed, when the conceptual structure of the problem was explained to them through appropriate diagramming techniques, they were able to solve it easily and, through supplementary work, better able to solve problems of the same type subsequently (Danesi, 2003a).

The problem was explained to the students as follows. First, they were told that it was necessary to set up a linear equation. To do so, they would need to represent the *timeline* and *quantity* schemata in a visual way. They were shown how to do this by an appropriate diagramming technique, such as the one above. This allowed the students to see the relation between the two ages in concrete terms, that is, in terms of the underlying *timeline* source domain. They were then able to understand much more concretely that representing John's and Mary's ages "four years from now" entailed moving John's and Mary's age-points "to the right by four" on the timeline, as shown on the above diagram.

It was surprising to find out how easily students comprehended the abstract notions involved after this kind of representational technique was explained to them. Finally, to set up an algebraic relation between the two ages, the students were told to think of the *time is a quantity* image schema. John's age is quantifiable as twice that of Mary's age. This was explained as a difference in the size of the *containers* that held the ages of each one. The final step was a straightforward one—namely, expressing in algebraic notation the overall "picture relation." From the diagrams, the students could see that John's age four years from now was (x + 9). They could also see that this was twice Mary's age at that time: (x + 4). This led them to construct the necessary equation easily. In a semiotic framework, story problems (and other abstract math texts) would be explained using CMT if the analysis of the problem requires it and if students are experiencing difficulties. The specific type of diagrammatic strategy that is appropriate in each specific case will vary, but the underlying principle remains the same—diagrams are used to concretize the implicit source domains used in a problem.

As mentioned throughout this book diagramming technique is, in strict semiotic terms, an iconic one; that is, it is one that models some concept in a concrete simulative way. The use of iconic teaching techniques has been a staple of enlightened educational methodology since time immemorial. CMT pedagogy simply extends this methodology so that it can take into account the metaphorical structure of many learning tasks, such as story problems in algebra.

The original pilot project investigating the CMT framework in teaching students to solve story problems more effectively was initiated in the fall of 1992 and terminated in the spring of 1999. It involved grade eight students. The results were published in 2003 (Danesi, 2003a). Over the seven years, several research assistants at the University of Toronto were trained in CMT and told simply to use whatever insights it afforded them to prepare learning materials (for example, illustrative problems, diagramming techniques, etc.) that could help students identified as having severe difficulties in solving story problems grasp the language of such problems through the use of iconicity. At the time, grade eight was the point in the Ontario math curriculum where the solving of story problems in algebra was introduced (a situation that has since changed). Five grade eight math teachers in three local schools participated in the project by helping the research team—consisting of myself and the research assistants—identify the subjects for the project. The students named by the teachers were given a two-hour test consisting of typical grade-eight level story problems. This allowed the team to ascertain that the students were indeed candidates for the project. Those who scored less than 20% accuracy were selected.

Seven students per year, ranging in age from 12 to 14, were chosen in this way. There were 49 students in all who were used as subjects over the seven year period. Each student met with a member of the research team for 15 minutes after school hours on a specified day of the week during the first term of school. As could be expected, new members were trained every year, as older members dropped out of the team for various reasons. The same problems used in the classroom were taught again to each student using the CMT framework as a guide: that is, the student was either shown how to represent the problem iconically or else taken through, step by step, the metaphorical layers that undergirded its formulation. The progress of each student was charted on a regular basis. At the end of the school year, the scores that each student obtained in story problem-solving in class were compiled and assessed. At the end of the project, the cumulative average score for the group of 49 students went up from 20% success to an accumulated average of 82% (standard deviation of 2.3%). Although such incredibly positive results could be attributed simply to the fact that the students received extra individualized attention, they nevertheless provided a clear basis for investigating the use of CMT and other semiotic frameworks further in math education. After all, the students had been given special attention prior to the intervention of the research team by the grade eight teachers themselves and by special educators in the schools, with only minimal success. In effect, the pilot project provided an empirical basis for conducting future work in this area.

As mentioned, a similar study was conducted in the fall of 2005. This time a class in the School of Continuing Studies at the University of Toronto was chosen, consisting of 30 adult students who defined themselves as math phobics, and all of whom identified story problems at the beginning of the course as being the most difficult ones for them to understand. The course was established as an experimental course in 2000 specifically to help math phobics acquire abilities in such areas as problem-solving. The course was taken primarily by learners to acquire the skills they needed to pass standardized math tests, such as the GMAT. A similar procedure was used as the one described above for the original study. So, without going into detail here, suffice it to say that the exact same framework and materials developed for the original CMT study were used by the instructor of the math phobic group, who was also trained in CMT beforehand.

At the beginning of the course the students found problems such as those listed below virtually impossible to solve. Some of these problems have already been discussed in this book.

Problems Utilized

[PROBLEM 4–19]

The sum of the units and tens digit of a number is 10. If the number is multiplied by 5 the product is 140. What is the number?

[PROBLEM 4–20]

The sum of the digits of a two-digit number is 9. If the digits are reversed the number is greater than the original number by 27. What was the original number?

[PROBLEM 4–21]

Georgina got back her grades on three tests, math, statistics, and chemistry. She got 82 on the math test, 74 on the statistics test, and 90 on the chemistry. Her physics test is tomorrow. In order to be the first of her class she needs to get an

85 average on the four tests. What grade must she get on her physics test tomorrow in order to achieve this average?

[PROBLEM 4–22]

In a famous driving competition, Debbie and David drove the same course at (average) speeds of 51 miles per hour and 54 miles per hour respectively. Debbie got to the end of the course in 40 minutes. How many minutes did it take David to get there?

[PROBLEM 4–23]

If it takes Amin 3 hours to complete a job of painting a room and it takes Hanna only 2 hours to paint the same room, how long would it take to paint the room if they work on it together (at the same time)?

[PROBLEM 4–24]

At a supermarket, apricots can be purchased on sale for $0.15 each and peaches for $0.20 each. At these bottom bargain rates, Kaara purchased a bag of 21 apples and pears together for $3.80. How many pears did he purchase?

[PROBLEM 4–25]

A mixture of 12 ounces of wine and water consists of 40% wine (by weight). How many ounces of water must be added to that mixture to produce a new mixture with only 25% wine in it?

[PROBLEM 4–26]

The ages of Robert and Mary together are 26. If you double their ages, Rob's age will exceed Mary's by 8 years. How old is each one?

[PROBLEM 4–27]

Three times Jamal's age, in years, is equal to eight times Becky's age, in years. And the difference between their ages is fifteen years. How old is each one?

> [PROBLEM 4–28]
> A movie chain sells children's tickets for half the adult price. If 5 adult tickets and 8 children's tickets cost a total of $27, what is the cost of an adult ticket?

At the end of the course, they were given similar problems to solve within an hour. The results matched those of the initial study. The subjects showed no difficulty whatsoever in representing such problems after learning how to decode them metaphorically in the course. Eight of the students solved all 10 of them successfully; eighteen solved 9 of them; two solved 8; and the remaining two solved 7. The median solving rate was, therefore, almost 9 (rounded off). Given that most of the students could hardly solve any of the above problems at the beginning of the course, the most logical conclusion that can be drawn is that the pedagogy used was effective. Needless to say, the result could in this case too be attributed simply to the fact that the students received specialized attention. But, at least in an indirect fashion, the project showed that the results anticipated by a CMT framework are replicable, and that age, background, and classroom situation seem to be non-factors. In effect, the two pilot projects have provided an empirical basis for conducting future work in this area.

Pedagogical Questions

The CMT framework described in this chapter for teaching how to solve story problems raises several important questions that will have to be addressed in future work. For example: Are the results it has produced in the studies described here truly replicable in other situations? Will they vary if a more controlled learning-teaching situation is examined, with experimental and control groups being compared? How can one distinguish between purely metaphorical concepts in learning tasks and other kinds of concepts? Is the same type of approach applicable to the teaching of other subject areas? And so on and so forth.

Pedagogically, CMT appears to provide relevant insights on how to render seemingly difficult abstract tasks into concrete ones by showing learners how to decode the language in which they are framed. As it turns out, this appears to be a relatively straightforward instructional task in itself, since it involves teacher and learners in an interactive way, as various applications of metaphor theories to education have shown (Cameron and Low 1999, Cameron 2003). Whereas the original group of grade eight students could solve only 20% of story problems before their interactive learning sessions, by the end their success rate went up to over 80% (on average). In the math phobic project, virtually no one could solve story problems at the start of the course. By the end, nearly 90% (= median of 9) could. The training in both cases consisted, essentially, in helping students translate ordinary language into appropriate image schemata through concretization. The difficulties the students had in representing problems initially were, arguably, difficulties in translation between two semiotic systems—the conceptual metaphor system inherent in story problem texts and corresponding algebraic representation of the texts. What now remains to be done is to verify the effectiveness of the framework experimentally (especially vis-à-vis control groups of learners) and more broadly to test its viability in the teaching of all kinds of math learning tasks. There is no reason to believe that anything but findings similar to the ones discussed here will emerge from future experimentation.

Having said this, much work remains to be done on the CMT framework. For example, not all mathematical texts have a rhetorical structure, and if they do, some might reveal other kinds of rhetorical properties, such as metonymic ones, as well as other kinds of meaning structures such as frames and prototypes (Clark and Cunningham 2006). Metonymy in particular plays a crucial role in text-construction. However, it is beyond the scope of the present chapter to deal with this issue. Suffice it to say that it surfaces in many kinds of math texts where the symbolism used involves a part-whole or whole-part system of reference. Thus, the first thing that must be done within a broader CMT framework is to carry out an analysis and classification of math texts, using CMT as an initial template. Obviously, some texts will show a higher degree of metaphoricity when compared to other kinds of texts. This implies, as a corollary, that math teachers will have to become more conversant with current metaphorical theories in linguistics and semiotics, in the same way that they have had to consider various psychological theories of learning in the past. As illustrated in this book, this should not constitute a problematic area given the simplicity with which semiotics approaches the task of analyzing meaning in all kinds of texts. Equipped with more precise notions of what constitutes a text and the conceptual structure it demonstrates, math teachers can easily anticipate learning difficulties and take appropriate pedagogical steps to help the learners overcome them.

The CMT framework is, essentially, applicable to the analysis component of the semiotic pedagogy of story problems, discussed above, since it identifies the image schemata that are present in such problems. This then suggests appropriate transference operations. The classification of story problems into thematic areas, as was done above, is the first step in fleshing out recurring schemata and in suggesting appropriate operations. For example, in age problems, the *time is a point on a timeline* schema implies addition or subtraction, whereas the *time is a quantity* schema implies multiplication or division. The same analysis can be conducted on any type of story problem. It must be emphasized, however, that CMT is not the "magic key" that will guarantee successful math learning in all situations. Like good piano teachers, experienced math teachers know all too well that each student is different, that each learning task presents its own kinds of problems and, thus, that patience is the operative word in all situations and contexts of learning. But frameworks such as CMT seem to have broad applicability, indicating that all kinds of learning tasks invariably involve signifying processes. It is for this reason that semiotically-based pedagogical proposals are beginning, in my view, to receive wider attention and stimulate much more interest than they have at any other time in the history of education.

In a semiotic approach to education, both the learner and the teacher will have to use their common sense in deciding *what* and *how much* to learn at various points in a course. In high school and university courses, the *what* is often established in advance. This does not, however, preclude the use of the semiotic approach described in this book. In such courses, the teacher and learner can use the curriculum provided as a basis for exploring the relation between representational systems and how to represent various tasks. Although this is not an ideal situation to be in—as learners and teachers constantly find out—it is something that should not discourage teachers from treating aspects of the given curriculum in ways that have been discussed in this book.

5
GEOMETRIC AND OTHER PROBLEMS

Mathematics would certainly have not come into existence if one had known from the beginning that there was in nature no exactly straight line, no actual circle, no absolute magnitude.

Friedrich Nietzsche (1844–1900)

Arithmetic, algebra, and geometry constitute (and have always constituted) the three basic subject areas of elementary mathematics education. Historically speaking, geometry is probably the one to have been taught first in the ancient schools and academies. The term *geometry*, which derives from the Ancient Greek *geo* ("earth") and *metrein* ("to measure"), describes what the ancient geometers actually did. They measured the size of fields, laid out accurate right angles for the corners of buildings, and carried out other such practical things. They used diagrams to represent their measurements and their layouts. Solving problems in geometry still requires, in fact, knowledge of how to draw and interpret diagrams correctly. And, of course, that is what semiotic pedagogy is essentially all about, namely converting math abstractions into concrete representamina that will display the necessary insights into problem content and structure. Geometry is the branch of mathematics that deals with the properties of points, lines, angles, curves, and shapes. It also deals with the relation between numbers and figures. In previous chapters, the number line was often used as a pedagogical device. Actually, the line is a perfect example of how arithmetic and geometry can be linked. It is a simple geometric figure that graphically represents the real numbers as a series of points distributed about a point arbitrarily designated as zero and in which the magnitude of each number is represented by the distance of the corresponding point from zero.

It is interesting to note that Plato saw geometry as an example of how humans go about seeking true knowledge. Whether or not this is true, it is certainly a fact that without the ability to solve problems in basic geometry one will never gain true mathematical knowledge. Geometry problems are miniature models of how logical reasoning unfolds and, thus, it comes as no surprise to find that they have always been central to math education, from Plato's times to the present day.

Needless to say, only the most basic of topics in geometry can be discussed in this chapter. The treatment in the first part will be limited to a schematic pedagogical discussion of Euclidean geometry. In addition, this chapter will look briefly at three other basic areas of mathematics, since these too are often part of an elementary math curriculum, namely, coordinate geometry, permutations and combinations, and elementary probability theory. Coordinate geometry involves the use of algebra to solve problems in geometry and vice versa; permutations and combinations (known together as *combinatorics*) is the study of the structural properties of arrangements; and probability theory is the mathematical study of chance. The latter branch was developed by the French mathematician Blaise Pascal (1623–1662) along with his countryman Pierre de Fermat (1601–1665) in order to help a high-living nobleman improve his chances of winning at gambling.

Geometry Problems

A problem in geometry involves measurement or determining relations in a generally deductive fashion. Probably in no other area of mathematics is the use of diagrams and appropriate naming more critical initially than in geometry. Let's start with lines and angles. A *straight line* is the shortest distance between two points, such as A and B below:

In actual fact, AB is a *line segment*, since the line on which A and B exist can be extended ad infinitum both to the left of A and to the right of B. So, the notation AB really denotes "line segment" or equivalently the "length" of the segment.

An *angle* is the figure formed by two lines that meet at a point common point. For example, lines AB and CB meet at the common point B, forming a "space" between them. That space is the angle represented by the notation $\angle ABC$ with the common point B, known as the *vertex*. The parts of the angle are named as shown:

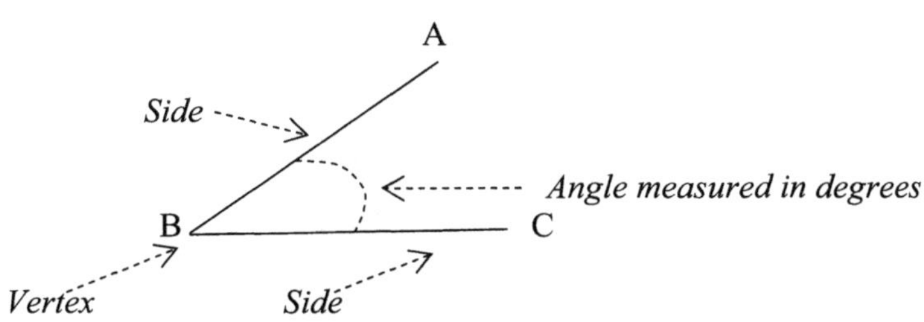

The number of degrees can be explained concretely as the rotation between the two sides that make up an angle. Degrees are shown with a superscript zero:

60^0 = sixty degrees
120^0 = one hundred and twenty degrees
...
n^0 = n degrees

There are many problems involving lines and angles. These usually require us to prove certain things. Take, for example, a problem that involves vertically-opposite angles. These are angles whose vertices touch, as a result of straight lines intersecting:

[PROBLEM 5–1]
Prove that the angles formed when two straight lines intersect are equal.

The idea is to show that such angles are necessarily equal under all circumstances, that is, no matter what their measure in degrees is. This is known as a *proof*. The early Greeks realized that observation and commonly-held notions did not guarantee understanding of truth (in human terms at least). To prove something as true, one must reason in a logical fashion, applying previous knowledge to the problem at hand. That is the core of PS in geometry.

We start, as always, with the technique of concretization by drawing a diagram that shows all the relevant features The two straight lines are labeled AB and CD, and two of the four vertically-opposite angles formed by their intersection can be labeled x and y. One of the angles between x and y is labeled z, as shown:

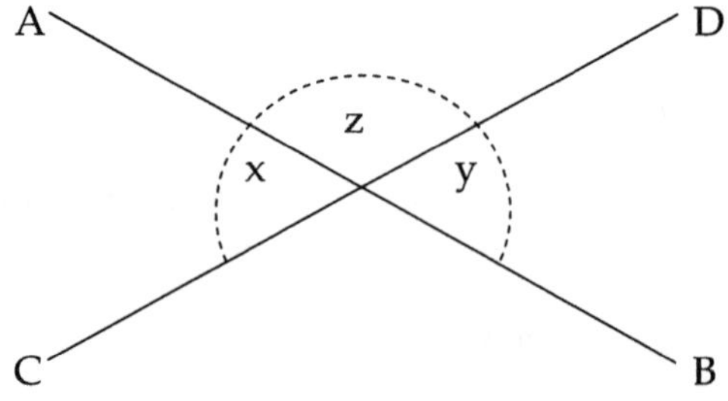

The objective of the problem is to demonstrate logically that x and y, being vertically-opposite angles, are always equal in measure. There are, of course, two other vertically-opposite angles formed by the intersection of the two lines, but they need not be considered here because the method of proof and the end result will be the same. The proof hinges on the established fact that a straight line is an angle of 180^0. We start by telling the student to consider line CD first. As a straight line, it is an angle of 180^0. Now, we point out that CD is composed of two smaller angles on the diagram, x and z. Logically, these two angles, being on the line CD, must add up to 180^0. This fact can, of course, be represented with an equation:

(1) $x + z = 180^0$

Now, we point out line AB too, is composed of two smaller angles on the diagram, y and z. These two angles also add up to 180^0: This fact can be similarly represented with an equation:

(2) $y + z = 180^0$

These two equations can be rewritten as follows:

(1) $x = 180^0 - z$ ← *Transposing* z *to the right side, changing its sign*
(2) $y = 180^0 - z$ ← *Transposing* z *to the right side, changing its sign*

Now, since things equal to the same thing are equal to each other we can deduce that x = y, since equation (1) shows that x is equal to $(180^0 - z)$, and equation (2) that y is equal to the same expression $(180^0 - z)$. It is not necessary to figure out what the value of the expression is. Whatever it is, the fact remains that x and y will each be equal to it. We can now conclude that any two vertically-opposite angles produced by the intersection of two straight lines are equal, because we did not assign a specific value to either angle. We have, in effect, proved a so-called proposition for the general case.

We can now introduce the concept of parallel lines, defined as straight lines that will never cross on the plane (on a sheet of flat paper, for instance), no matter how far they are extended. Such lines form a perfect binary opposition with intersecting lines, of course. And this is why they should follow the previous presentation. The symbol for parallel is $||$: AB $||$ CD means that AB and CD are parallel to each other. A line that crosses other lines is called a *transversal*. If a transversal (EF below) crosses two parallel lines (AB and CD) the following angle relations materialize, because of the previous proof (problem [5–1]:

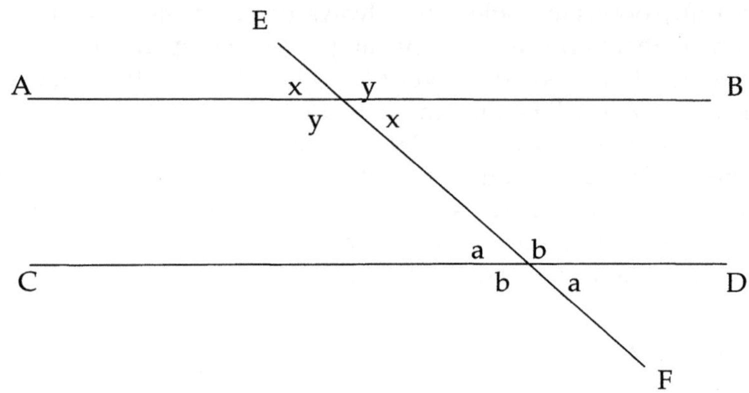

Again, it should be emphasized to students that the angles are shown equal each other because they are produced by intersecting line segments. Now, we can introduce an appropriate problem:

> [PROBLEM 5–2]
>
> If a transversal crosses two parallel lines it will make the angles on opposite sides of the transversal (within the two parallel lines) equal. In terms of the diagram above this translates as follows: Prove that x = a and y = b.

Let's start by proving x = a. First, we label the points of intersection as shown:

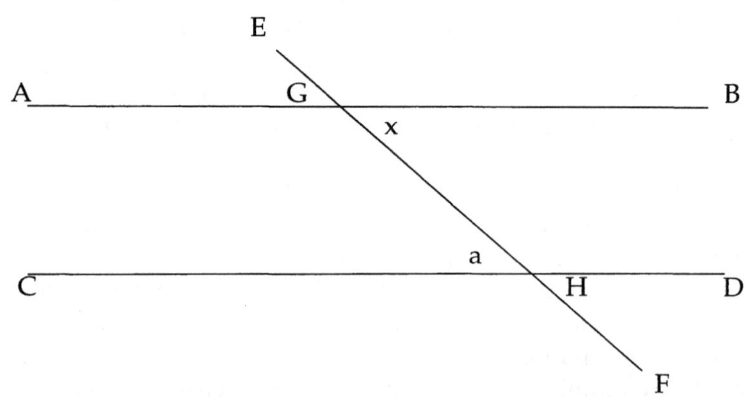

In order to prove what we are supposed to, something must clearly be done to the diagram. We can ask students at this point what this could be. Ultimately, either the students or the teacher will suggest that dropping a perpendicular line from the point G on to the line CD is the additional feature that must be added to the diagram. Why (the students might ask)? The reason is that this produces a right-angled triangle, allowing us to envision a potential use of the Pythagorean Theorem:

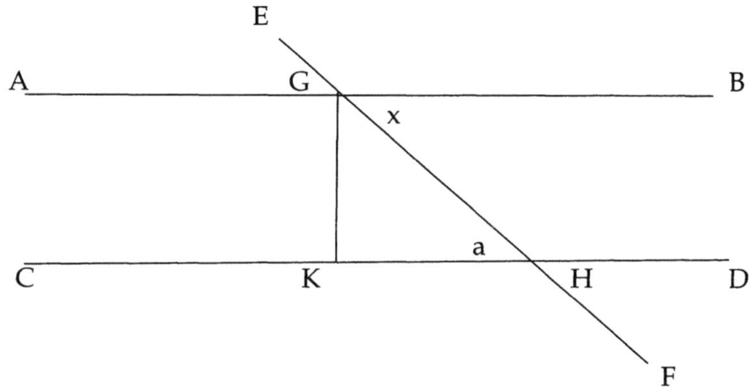

Next, we label other relevant angles on the diagram as shown:

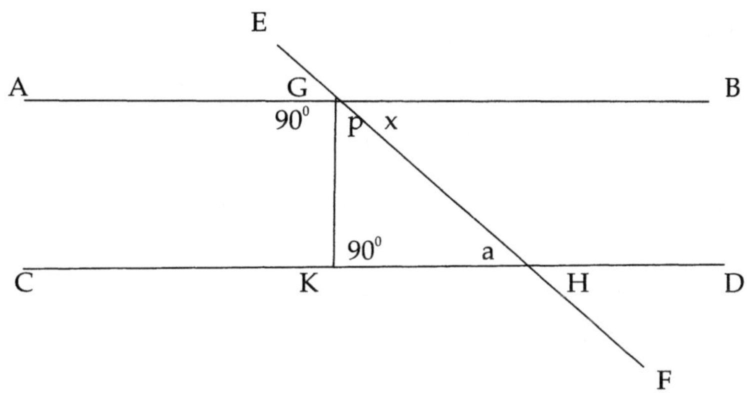

We know that the sum of the angles in a straight line is 180^0. Consider line AB. It is divided into three angles, 90^0, p, and x. So, these add up to 180^0:

$x + p + 90^0 = 180^0$

$x + p = 180^0 - 90^0$ ← *Transposing 90^0, changing its sign*

(1) $x + p = 90^0$ ← *Simplifying the right side*

Now, we tell students to consider triangle GKH. We know that the angles of a triangle add up to 180^0. The angles of GKH are a, p, and 90^0. So, these add up to 180^0:

$a + p + 90^0 = 180^0$
$a + p = 180^0 - 90^0$ ← *Transposing 90^0, changing the sign*
(2) $a + p = 90^0$ ← *Simplifying the right side*

Now, we tell students to look at equations (1) and (2) and they will see that $(x + p)$ and $(a + p)$ both equal 90^0. So, they equal each other:

$a + p = x + p$

Since p is common to both sides, it can be subtracted from both, leaving:

$a = x$

This proves what the problem asked us to prove (in part). The same method of proof can be used to prove that $b = y$. There is no need to go through it here. We have, actually, proved a very important proposition. We indicate that a proposition is a generalization ensuing from a problem solved in general. Since, we gave no values to the angles a and x, the proof of problem [5–2] is a generic one and, thus, is called a proposition.

Now, we can write a series of problems based on this proposition. Following is an example of such a problem, which the student can now easily handle:

[PROBLEM 5–3]

Given the diagram below, what is the value of x?

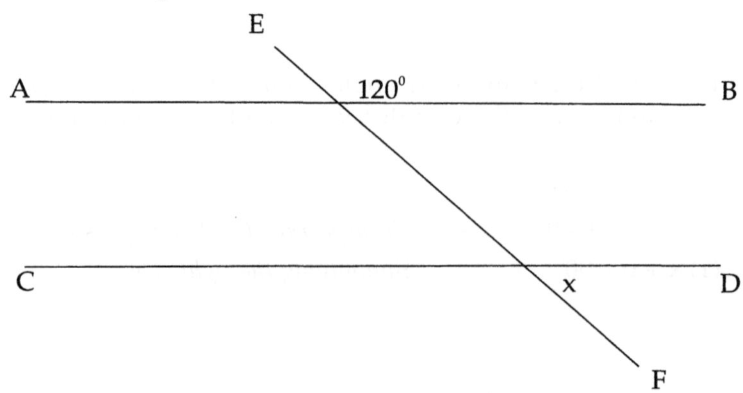

We know that a straight angle is 180°, so the supplementary angle of 120° at the top is 60°, since 120° + 60° = 180°. Let's mark this on the diagram:

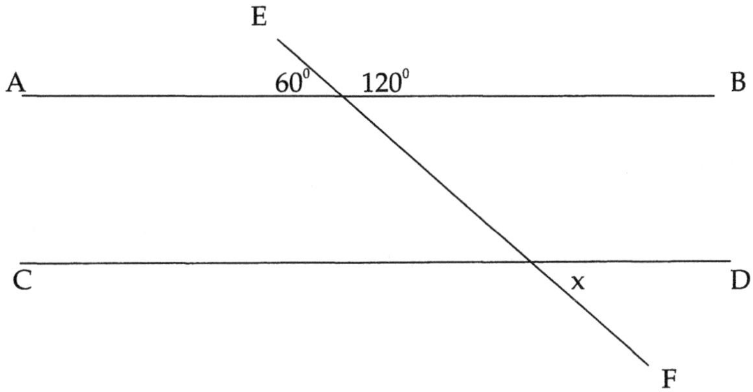

Since we know that vertically-opposite angles are equal, we can thus mark the diagram up as follows:

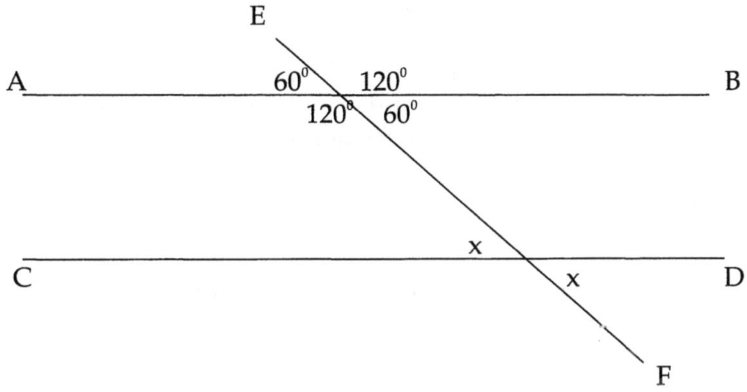

We also know that angles on the opposite sides of a transversal crossing two parallel lines are equal. So, we can now state that x = 60° because x is an opposite angle of 60°.

The cognitive process involved in solving geometry problems such as this one is a deductive one. It can be characterized as the process of showing that something is necessarily so on the basis of previously-established facts. Many problems in geometry involve deduction. However, some entail the use of induction, or arriving at some generalization through the analysis of the pattern inherent in specific facts. Here is a basic problem involving inductive reasoning:

[PROBLEM 5–4]
Derive a formula for determining the number of degrees in the angles of a polygon.

With all inductive PS we start with the simplest possible case. So, let's consider a triangle first—the polygon with the least number of sides—three. As the students know by now, the sum of the angles in a triangle is 180^0. We tell students to mark this fact down in a chart. We now go on to consider a quadrilateral (the next case, since it is a polygon with four sides). We start by drawing one, with a diagonal through it. Why (the student may ask)? The reason is that it creates two triangles, indicating that this structural result might provide a clue to the overall solution.

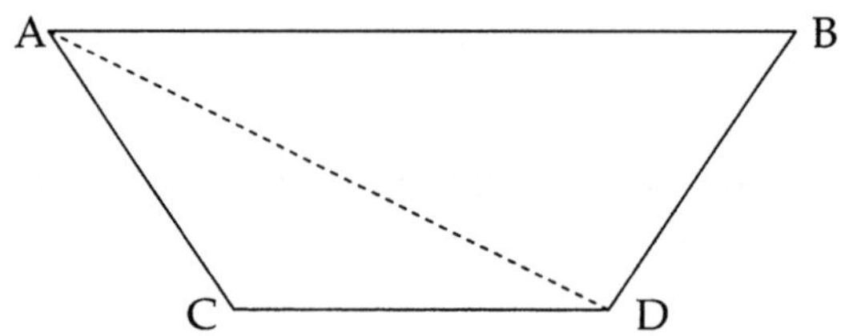

Two triangles can also be produced by joining BC instead. By doing this we have discovered that the sum of the angles in the quadrilateral is equivalent to the sum of the angles in two triangles, namely:

$$180^0 + 180^0 = 360^0$$

Next, we consider the case of a pentagon (a five-sided figure). ABCDE below is one such figure:

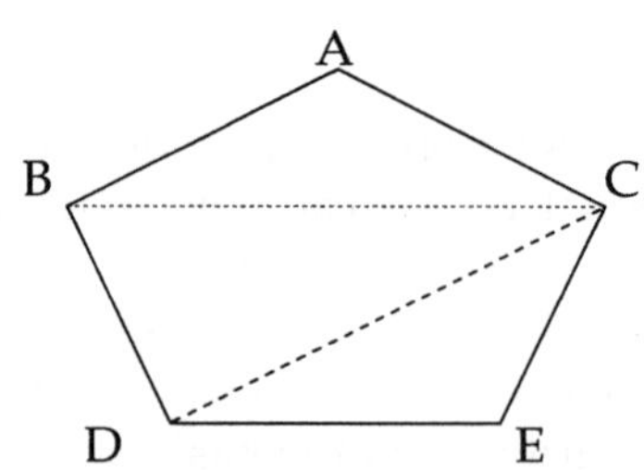

The pentagon can be divided into three triangles, with two diagonals, as shown above (triangle ABC, triangle BCD, and triangle CDE). There are other ways to partition the pentagon, but the result will always be the same—three triangles. We have now discovered another simple fact—the sum of the angles in a pentagon is equivalent to the sum of the angles in three triangles:

$$180^0 + 180^0 + 180^0 = 540^0$$

Continuing on in this way, it can be shown just as easily that the number of angles in a hexagon is equal to the sum of the angles in four triangles (with three diagonals), in a heptagon to the sum of the angles in five triangles (four diagonals), and so on. Let's now generalize what we have apparently discovered. It would appear that the number of triangles that can be drawn in any polygon is "two less" than the number of sides that define the polygon. For example, in a quadrilateral we can draw two triangles, which is "two less" than the number of its sides (4), or (4 - 2); in a pentagon, we can draw three triangles, which is, again, "two less" than the number of its sides (5), or (5 - 2); and so on. In the case of a triangle, this rule also applies, since we can draw in it one and only one triangle (itself). This also is "two less" than the number of its sides (3), or (3 - 2). In an n-gon, therefore, we can draw (n - 2) triangles. We should now use a chart to summarize these discoveries:

Number of sides in the polygon	*Number of triangles*
3 (= triangle)	(3 - 2) = 1 triangle
4 (= quadrilateral)	(4 - 2) = 2 triangles
5 (= pentagon)	(5 - 2) = 3 triangles
6 (= hexagon)	(6 - 2) = 4 triangles
7 (= heptagon)	(7 - 2) = 5 triangles
...	...
n (= n-gon)	(n - 2) triangles

Since we know that there are 180^0 in a triangle, then there will be 2×180^0 in a quadrilateral, 3×180^0 in a pentagon, and so on. In an n-gon, therefore, there will be $(n - 2)\ 180^0$:

Sides	*Triangles*	*Sum of Degrees*
3	(3 - 2) = 1	$180^0 \times 1 = 180^0$
4	(4 - 2) = 2	$180^0 \times 2 = 360^0$
5	(5 - 2) = 3	$180^0 \times 3 = 540^0$
6	(6 - 2) = 4	$180^0 \times 4 = 720^0$
7	(7 - 2) = 5	$180^0 \times 5 = 900^0$
...	...	...
n	(n - 2)	$180^0 \times (n - 2) = (n - 2)\ 180^0$

Now, with this formula in hand, we can determine the number of degrees in any polygon. An appropriate problem is the following one:

[PROBLEM 5–5]
How many degrees are there in an octagon?

In the case of an octagon, n = 8. Plugging this value in our formula will yield the number of degrees in an octagon:

$$(n - 2)\,180^0 = (8 - 2)\,180^0 = 6 \times 180^0 = 1080^0$$

A polygon with all sides the same length is known as a *regular polygon.* The angles in such a polygon are all equal. This property makes it easy to figure out what the size of each angle is. For example, in a regular polygon, we know that the sum of all five angles is 540^0. Since each angle is the same, we divide $540^0 \div 5 = 108^0$. This is the size of each angle in a regular pentagon.

At this point it is appropriate to deal with arguably the most important proposition or theorem of geometry, the Pythagorean Theorem, which states that the square on the hypotenuse of a right-angled triangle is equal to the sum of the squares on the other two sides, or $c^2 = a^2 + b^2$:

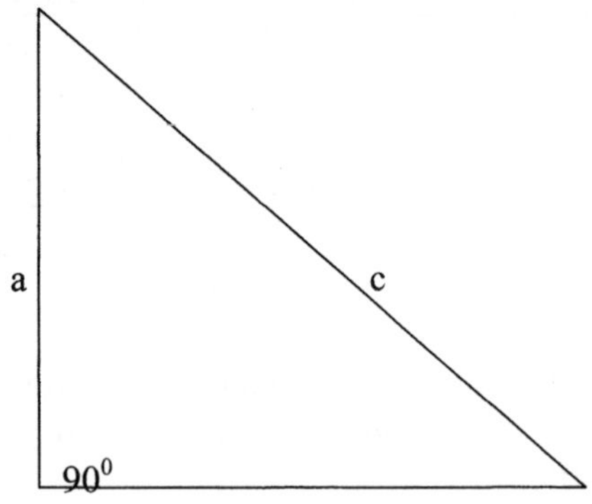

In the equation $c^2 = a^2 + b^2$, c is the length of the hypotenuse, and a and b the lengths of the other two sides, also known as *legs*. We can show the applications of this theorem concretely by exemplifying it in a concrete way. For example, if the length of a is 3 and the length of b is 4 then the length of c is 5:

$$c^2 = 3^2 + 4^2$$

$c^2 = 9 + 16$ ← *Squaring the digits*
$c^2 = 25$ ← *Simplifying the right side*
$c = \sqrt{25}$ ← *Taking the square root of both sides*
$c = 5$

The Pythagorean Theorem is used to solve a host of problems, as we have seen. Here's a very simple one that can be used propitiously in the classroom at this point. Although it is a straightforward closed problem, it is nevertheless important to get students to solve it for the sake of initial practice:

[PROBLEM 5–6]
In the right triangle below, you are given the lengths of one leg and the hypotenuse. Figure out the length of the second leg.

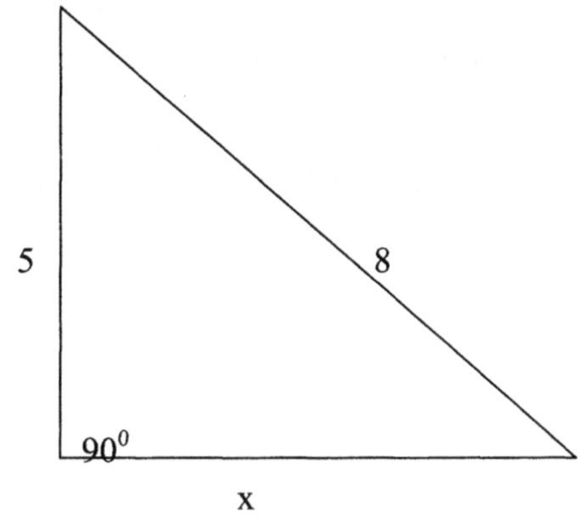

From the general statement of the Pythagorean Theorem, we know that:

$8^2 = 5^2 + x^2$
$x^2 + 5^2 = 8^2$ ← *Rearranging*
$x^2 = 8^2 - 5^2$ ← *Transposing 5^2, changing its sign*
$x^2 = 64 - 25$ ← *Expanding*
$x^2 = 39$ ← *Simplifying*
$x = \sqrt{39}$ ← *Taking the square root of both sides*
$x = 6.24$

As an historical aside, it can be mentioned that many cultures throughout the ancient world knew of this relation. The archeological discovery of a Babylonian method for finding the diagonal of a square, suggests that the theorem was known 1,000 years before Pythagoras (Musser, Burger, and Peterson 2006: 763). Pythagoras was probably the first one to prove that it holds for all right-angled triangles, even though he left no written version of it. There are hundreds of proofs for the theorem. Among those who provided proofs are the great Renaissance painter and scientist Leonardo da Vinci (1452–1519) and the American James Garfield (1831–1881), who was sworn in as the twentieth president of the United States on March 4, 1881. Garfield published his proof in an 1876 issue of the *New England Journal of Education* while still a member of the House of Representatives.

Another important topic in elementary geometry is that of area. *Area* is the measure of the amount of surface enclosed by a polygon. It is measured in *square units.* In the case of a triangle the area is defined as half the length of a base (b) multiplied by the corresponding height (h): 1/2 bh. The reason behind this formula can be shown easily. The area of a quadrilateral, such as a rectangle (a typical floor layout) is, as students know already from everyday life, the length multiplied by the width. If we replace the word *length* with *base* and *width* with *height,* the formula above indicates that the area of a triangle is half that of a rectangle. And, of course, this is correct because any quadrilateral can be cut up into two triangles, as we have seen above. Take, for instance, the rectangle below. As the student can see, a diagonal will divide it into two triangles, each of which is half the area of the rectangle:

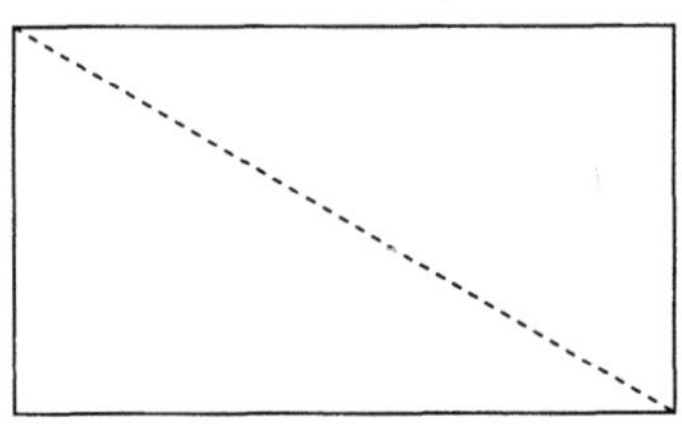

We point out that that the height of a triangle must be a perpendicular. So, in the case of a right triangle, one of the legs is also the height. In the case of other types of triangle, it will be necessary to drop a perpendicular from a vertex:

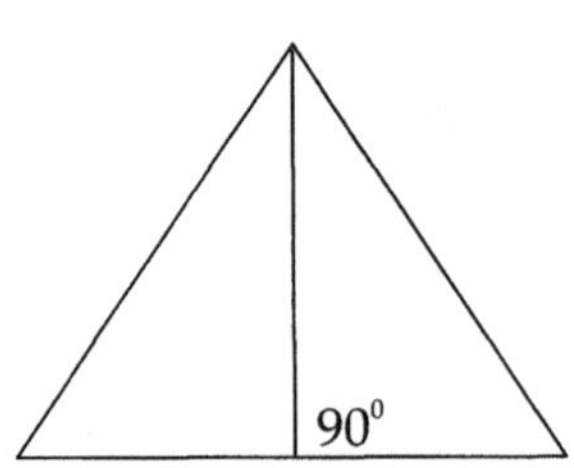

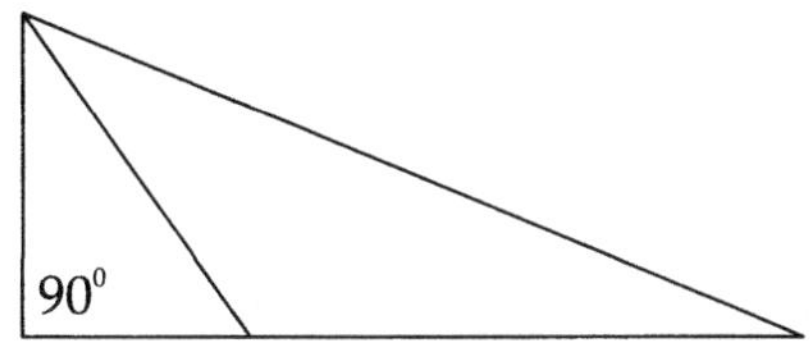

Closed problems in area can easily be drawn up. Here are a two, which are self-explanatory:

> [PROBLEM 5–7]
> What is the area of an acute triangle with base 2 and height 5?

> [PROBLEM 5–8]
> What is the area of a polygon when it is known that it is equal to the area of a square of length 6?

Without going further into other elementary geometrical topics (circles, solids, etc.) suffice it to say that the pedagogy remains the same even though the structures change. The use of semiotic techniques is, actually, consistent with the well-known work of the Van Hieles (summarized in Van Hiele 1984). In the late 1950s in the Netherlands, two math teachers, Pierre van Hiele and Dieke van Hiele-Geldof, noticed that students learning basic geometry tended to do so in a sequence of levels, which they named as follows:

- Level 0 (recognition): The learner recognizes certain geometric shapes as consistent, regardless of size, orientation, etc. Semiotically, the learner recognizes the difference between structure and form (a triangle has the same structure, regardless of the particular form it takes on).

- Level 1 (analysis): The learner starts to focus on the component parts of figures (sides of a triangle, diameter of a circle, etc.) and their relations to each other (equality, inequality, etc.). As we saw in the previous chapter, analysis entails learning to focus on the properties of representations and what they stand for.

- Level 2 (relationships): The learner starts to make relations among figures recognizing, for example, that squares and rectangles are quadrilaterals. In semiotic terms, the student develops a sense of system.

- Level 3 (deduction): This is the level at which learners start to work consciously with proofs and abstract processes. Semiotically, the learner is engaging in generalization and thus symbolic processes.

- Level 4 (axiomatics): At this level, learners no longer need diagrammatic or other representational aids, being able to formulate general axioms about the nature of geometrical figures and relations.

Coordinate Geometry Problems

The fields of arithmetic, algebra, and geometry are highly interrelated—one assumes knowledge of the other. As discussed in chapter 3, algebra is general arithmetic and geometry is, in many ways, visual arithmetic. This was known to the ancient mathematicians. However, the formal amalgamation of arithmetic and geometry had to await the work of French mathematician and philosopher René Descartes. He called his amalgamation *analytic geometry*. The basic notion in analytic geometry is that of intersecting number lines. A number line, as mentioned, is itself a rudimentary geometric representation that shows the continuity between positive and negative numbers and a one-to-one correspondence between a specific number and a specific point on the line.

Descartes simply drew two number lines intersecting at right angles. The horizontal line is called the x-axis, the vertical one the y-axis, and their point of intersection the origin. This system of two perpendicular intersecting number lines is called eponymously the *Cartesian plane*:

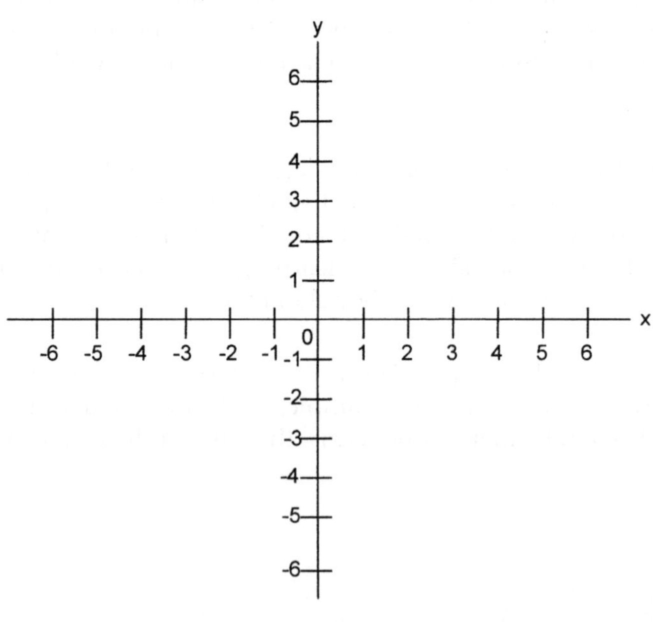

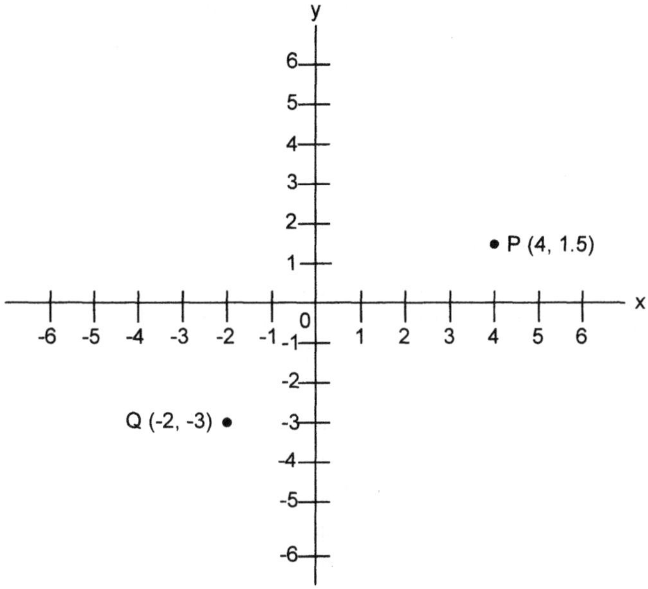

This simple diagram is also called a *coordinate system*, because it allows the plane to be mapped as a system of points that are determined by their positions in relation to the two axes. For example, the paired coordinates for point P in the figure above are (4, 1.5) and for Q they are (-2, -3). This means that point P is four units to the right of the y-axis, and one and a half units directly above the x-axis and that Q is two units to the left of the y-axis and three down from the x-axis:

Each point in the plane can be identified by such ordered pairs of coordinates. The first one is the x-coordinate, expressing the distance to the left or right of the y-axis. If it is positive it is to the right; if it is negative it is to the left. The second one is the y-coordinate, expressing the distance above or below the x-axis. If it is positive, it is above; if it is negative, it is below. The point of intersection of the two axes, O, is called the origin. Its coordinates are, of course, (0,0). Generally, points can be represented as follows:

$P_1\ (x_1, y_1)$ ← *A point with coordinates x_1 and y_1*
$P_2\ (x_2, y_2)$ ← *A point with coordinates x_2 and y_2*
$P_3\ (x_3, y_3)$ ← *A point with coordinates x_3 and y_3*
$P_4\ (x_4, y_4)$ ← *A point with coordinates x_4 and y_4*
...
$P_n\ (x_n, y_n)$ ← *General formula for a point in the plane*

We can now introduce an exploratory problem that requires students to use their previous knowledge of geometry applied to the Cartesian plane. As with all exploratory problems, this constitutes an effective pedagogical tool because it allows students to interrelate theory and PS tasks on their own:

> [PROBLEM 5–9]
>
> Can you find the distance between points P and Q in the Cartesian plane above?

To find the distance between the two points P (4, 1.5) and Q (-2, -3), we join them together to produce the line segment PQ. We can then draw a right triangle with PQ as hypotenuse, and easily figure out the coordinates of R. These are (4, -3):

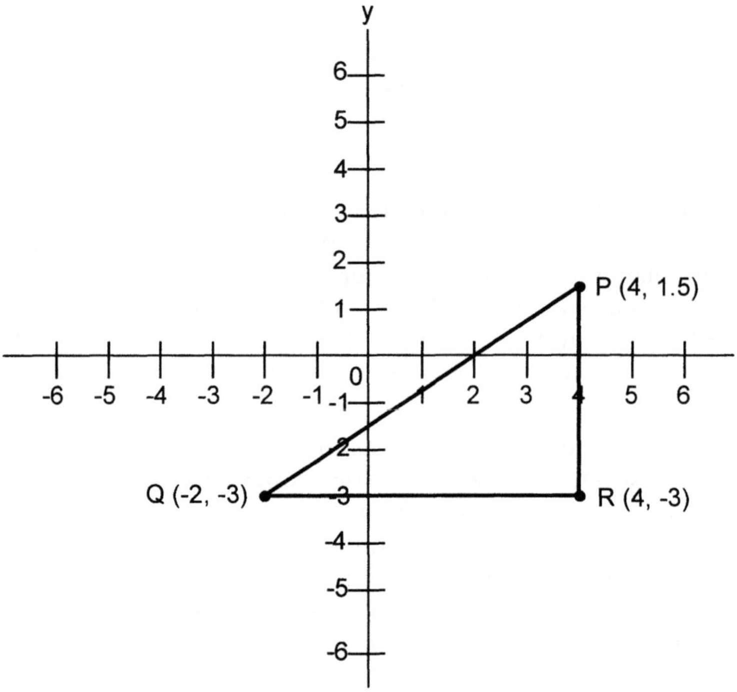

The lengths of the legs of the right triangle can be easily determined by looking at the two axes. We tell students to ignore the signs of the coordinates, considering only their absolute values. PR is 1.5 units above the x-axis and 3 below it. In total it is 1.5 + 3 = 4.5 units in length. QR is two units to the left of the y-axis and four units to its right. In total it is 2 + 4 = 6 units in length. PQ is the hypotenuse of triangle PQR. So, applying the Pythagorean Theorem, we get:

$PQ^2 = PR^2 + QR^2$ ← *Pythagorean Theorem*
$PQ^2 = 4.5^2 + 6^2$ ← *Plugging in the appropriate values*
$PQ^2 = 20.25 + 36 = 56.25$ ← *Simplifying*
$PQ = \sqrt{56.25} = 7.5$ ← *Taking the square root of both sides*

Now, we can ask students if it is possible to derive a general formula for determining the length of a line in the Cartesian plane. This is a problem in formalization, that is, in reasoning from the particular to the general:

> [PROBLEM 5–10]
> Given what you now know, can you derive a general formula for the length of a line in the plane?

We note that:

- The line can be made to constitute the hypotenuse of a right-angled triangle (as shown above).

- The length of the two legs can be determined by adding the absolute values of the corresponding coordinates of the end points, namely $|x_1 + x_2|$ for one leg and $|y_1 + y_2|$ for the other.

So, the general formula for the length of any line (L) is:

$$L^2 = (|x_1| + |x_2|)^2 + (|y_1| + |y_2|)^2$$
$$L = \sqrt{(|x_1| + |x_2|)^2 + (|y_1| + |y_2|)^2}$$

In the example above, $x_1 = 4$ so $|4| = 4$, $x_2 = -2$ so $|-2| = 2$, $y_1 = 1.5$ so $|1.5| = 1.5$, and $y_2 = -3$ so $|-3| = 3$. If we plug these values into the formula, we get:

$$L = \sqrt{(|x_1| + |x_2|)^2 + (|y_1| + |y_2|)^2}$$
$$L = \sqrt{(4 + 2)^2 + (1.5 + 3)^2}$$
$$L = \sqrt{(6)^2 + (4.5)^2}$$
$$L = \sqrt{36 + 20.25}$$
$$L = \sqrt{56.25} = 7.5$$

We can now revisit the concept that a straight line, such as PQ , is a *linear equation*. This, as students know by now, is an equation with its variable to the first power. The general form of the linear equation is as follows:

$$y = mx + b$$

In this formula m and b are constants—m is called the *slope* of the line because it indicates the degree to which it leans to the left or to the right. A line with 0 slope would be the y-axis. Any slope to the left of this axis is negative and to the right positive. The constant b is called the *y-intercept*, since it tells us where the line will cross the y-axis. The slope of a line passing through any two points, P (x_1, y_1) and Q (x_2, y_2), is defined as follows:

$$(y_1 - y_2) / (x_1 - x_2), \text{ if } (x_1 \neq x_2)$$

For example, in the case of PQ above where the coordinates of P are (4, 1.5) and of Q (-2, -3), the slope can be easily determined as being 0.75, since $y_1 = 1.5$, $y_2 = -3$, $x_1 = 4$, and $x_2 = -2$:

$$\begin{aligned} &(y_1 - y_2) / (x_1 - x_2) = \\ &[1.5 - (-3)] / [4 - (-2)] = \\ &[1.5 + 3] / [4 + 2] = \\ &4.5 / 6 = 0.75 \end{aligned}$$

A positive slope indicates, as mentioned, "slope to the right." And, indeed, PQ slopes to the right, as the formula predicts. The y-intercept is the distance from the origin to the point where the straight line intersects the y-axis. In our graph of PQ above that point is -1.5. Now, that we know the slope (0.75) and y-intercept (-1.5), we can develop the appropriate equation that describes the line PQ:

$y = mx + b$ ← *Formula for a linear equation*
$y = 0.75x - 1.5$ ← *Plugging in the appropriate values*

As this example shows, analytic geometry allows us to determine what type of equation describes a geometrical figure and vice versa. In other words, it is a code that generates algebraic models (equations) of geometric figures. Equations in which the variable is to the power of two describe figures such as circles, ellipses, and the like. The next problem is exploratory in this regard:

[PROBLEM 5–11]

For what figure does the equation $x^2 + y^2 = 25$ stand for?

The equation turns out to be the equation of a circle. To get the students to see this, we assign values to one of the variables and solve for the other one in the usual fashion:

Let $x = 0$

$x^2 + y^2 = 25$
$0 + y^2 = 25$
$y = \pm 5$ ← *Taking the square root of both sides*

Let $x = 4$

$x^2 + y^2 = 25$
$4^2 + y^2 = 25$
$16 + y^2 = 25$
$y^2 = 25 - 16$ ← *Transposing 16, changing its sign*
$y^2 = 9$ ← *Simplifying*
$y = \pm 3$ ← *Taking the square root of both sides*

Continuing on in this way and then plotting the values, the figure of a circle emerges on the Cartesian plane:

Values of x	0	0	+4	+4	-4	-4	...
Values of y	+5	-5	+3	-3	+3	-3	...

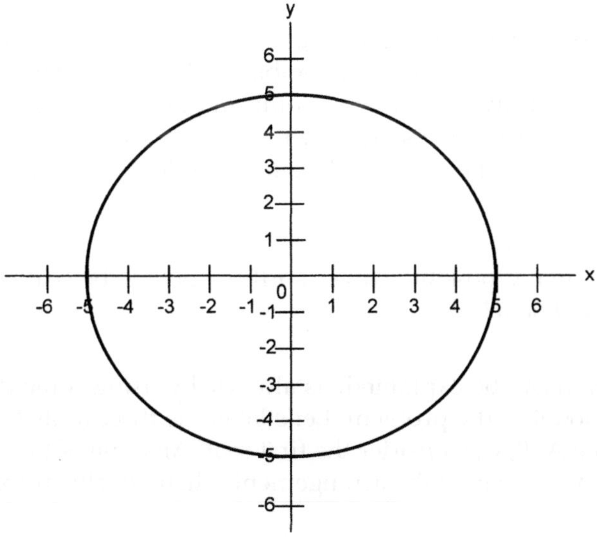

The technique used to solve this problem introduces an important concept—that of a *function*, or the relationship that holds between variables. One of the variables is referred to as the *independent variable*. In our example, it is x, because it is the variable to which we assign any value we want. The other variable is called the *dependent variable*, because it is the variable whose values correspond (or are dependent upon) the values assigned to the independent variable. In our example, it is y. The term *function* was first used by the German mathematician Gottfried Wilhelm Leibniz (1646–1716).

In our example, the function is determined as follows:

$$x^2 + y^2 = 25$$

$$y^2 = 25 - x^2 \quad \leftarrow \textit{Transposing } x^2\textit{, changing its sign}$$

$$y = \sqrt{(25 - x^2)} \quad \leftarrow \textit{Taking the square root}$$

The formula for the function is written with the symbol f(x). So:

$$f(x) = \sqrt{(25 - x^2)}$$

This notation is due to the Swiss mathematician Leonhard Euler. To summarize, it is crucial for students to grasp that coordinate geometry is a code for amalgamating algebra and geometry. If this is not emphasized, then learning difficulties tend to emerge, as various studies have shown (for example, Adams 1993) and as teaching experience often brings out.

Problems in Combinatorics

The study of how things can be arranged under certain specified conditions is called *combinatorics*. Semiotically, combinatorics is a syntagmatic counting operation; that is, it involves how to count in pairs, in triplets, and in other ordered or combinatory ways. The two main types of problems in this domain are called *permutations* and *combinations*. Following is a typical one:

> [PROBLEM 5–12]
> In how many possible ways can three people be seated on a three-seat boat?

The solution can be explained, as always, by using some diagrammatic or schematic model of the problem. Let's label the three seats (1), (2), (3) and the three people A, B, C. Consider the first seat. Any one of the three persons could sit in it. We can show the arrangement schematically as follows:

Seat (1)	Seat (2)	Seat (3)
↓	↓	↓
A	__	__
B	__	__
C	__	__

Now with A in (1) either B or C can sit in (2); with B in (1), either A or C can sit in (2); and with C in (1), either A or B can sit in (2). The possible seating arrangements so far can be shown schematically as follows:

Seat (1)		Seat (2)	Outcome	Seat (3)
↓		↓		↓
A	with	B	AB	__
A	with	C	AC	__
B	with	A	BA	__
B	with	C	BC	__
C	with	A	CA	__
C	with	B	CB	__

We note that the total number of potential pairings so far (shown in the Outcome column above)—AB, AC, BA, BC, CA, CB—is $3 \times 2 = 6$. This expresses in arithmetical form the fact that for each of the 3 people in seat (1), 2 others can occupy seat (2), producing a total of 6 pairs.

Now, for each of these 6 pairs, there is 1 choice left for occupying seat (3). For example, with the pair AB in seats (1) and (2) respectively, only C is left to occupy seat (3); with AC in seats (1) and (2) respectively, only B is left to occupy seat (3); and so on. Here is a schematic summary of all the possible arrangements of the three people on the boat:

Seat (1)		Seat (2)	Outcome		Seat (3)	Final Outcome
↓		↓			↓	
A	with	B	AB	with	C	ABC
A	with	C	AC	with	B	ACB
B	with	A	BA	with	C	BAC
B	with	C	BC	with	A	BCA
C	with	A	CA	with	B	CAB
C	with	B	CB	with	A	CBA

We point out, again that the total number of potential pairings (shown in the Final Outcome column above)—ABC, ACB, BAC, BCA, CAB, CBA—is $3 \times 2 \times 1 = 6$. As before, this expresses in arithmetical form the fact that for each of the 3 people who can occupy seat (1), 2 can occupy seat (2), and 1 seat (3), producing a total of 6 triplet arrangements. The solution to the problem can now be used to present a new counting notion called a *permutation*. A permutation can now be defined, more precisely, as an ordered arrangement of the elements of a set. The set in this case was {A, B, C}. We can now use PS in this domain to get the student to explore structural characteristics of permutations and, thus, generalize them formulaically. The use of PS in this way is called *heuristics*. It can be defined as the pedagogical technique of using PS as an exploratory guide to further learning:

> [PROBLEM 5–13]
> Consider a variant of the previous problem, namely a four-seat boat. In how many ways can four people be seated in it? Then, move on to a five-seat version. From such problems, can you derive a general statement or formula for solving all problems of this type?

Consider the four-seat version. In this case, there will $4 \times 3 \times 2 \times 1 = 24$ possible arrangements. The first digit, 4, refers to the fact that any one of four people—call them A, B, C, D—can occupy seat (1). The second digit 3 tells us that for each of the four in seat (1) there are three ways in which seat (2) can be occupied. This produces $4 \times 3 = 12$ permutations for the first two seats. The third digit tells us that for each of the 12 seating arrangements, there are two ways in which seat (3) can be occupied, for a total of $4 \times 3 \times 2 = 24$ permutations. Finally, the fourth digit indicates that for each of the 24 seating arrangements so far, there is only one way in which seat (4) can be occupied, for a total of $4 \times 3 \times 2 \times 1 = 24$ permutations.

If there were five places to be occupied on the boat and five people to fill them, there would be $5 \times 4 \times 3 \times 2 \times 1$ possible permutations (as students can reason for themselves); if there were six places and six people, there would be $6 \times 5 \times 4 \times 3 \times 2 \times 1$ permutations; and so on. We now ask the students if they detect a pattern, algebraically speaking. If there were n places to be occupied, and n people to fill them, there would be $n \times (n - 1) \times (n - 2) \times \ldots 1$ permutations. This is known as a *factorial*, and is represented with the symbol n!:

$$n! = n \times (n - 1) \times (n - 2) \times \ldots 1$$

This formula generalizes the fact that the first position can be filled with n objects, the second position with one less than the total number of objects (n - 1), the third with two less (n - 2), and so on down to a single possibility for the last position:

$$1! = 1$$
$$2! = 2 \times 1$$
$$3! = 3 \times 2 \times 1$$
$$4! = 4 \times 3 \times 2 \times 1$$
$$5! = 5 \times 4 \times 3 \times 2 \times 1$$
$$\ldots$$
$$n! = n \times (n - 1) \times (n - 2) \times \ldots 1$$

It is interesting to note that heuristic investigation, and the kinds of discoveries it leads to, brings out the cognitive dynamic that exists between signs, discovery, and inferential processes—discussed briefly and anecdotally in previous chapters. This relation will be discussed again below as a conclusion to this chapter.

Now, let's consider a situation in which we have more objects than positions in which to place them. An appropriate problem in this case, would be the following one:

> [PROBLEM 5–14]
>
> Suppose you had to choose one person to be president, one vice-president, and one secretary from a committee of 10 people. In how many ways could you fill those three positions?

Of course, 10 people can be selected for the position of president; for each of these any one of the remaining 9 can be selected for the vice-presidency; and for each of the previously selected pairs, we can choose from among 8 people to fill the remaining position of secretary. So, the number of total permutations in this case is 10 × 9 × 8 = 720.

Can this solution be generalized? We note that the answer of 10 × 9 × 8 is, in effect, 10! with the last seven factors canceled from it:

$$10 \times 9 \times 8 \times (\cancel{7 \times 6 \times 5 \times 4 \times 3 \times 2 \times 1})$$

This result can also be produced by dividing 7! (= 7 × 6 × 5 × 4 × 3 × 2 × 1) into 10! as shown below:

$$\frac{10!}{7!} = \frac{10 \times 9 \times 8 \times \cancel{(7 \times 6 \times 5 \times 4 \times 3 \times 2 \times 1)}}{\cancel{(7 \times 6 \times 5 \times 4 \times 3 \times 2 \times 1)}} = 10 \times 9 \times 8 = 720$$

We can now point out that the denominator 7! in the fraction above can cleverly be rewritten as (10 - 3)! and that the 3 in the parenthesis stands for the number of positions to be filled. This resourceful bit of rewriting provides the relevant insight. In general, if the numerator is n! then the denominator will be (n - r)! where r stands for the number of positions to be filled:

$$\frac{n!}{(n-r)!}$$

This formula now allows us to solve any problem that requires us to put n objects in r positions easily. Another way to phrase it is that the formula allows us to permute n objects taken r at a time. Again, we arrived at this through extrapolation, rather than straightforward application of previous knowledge.

Let's look at another type of arrangement pattern. Here is the relevant heuristic problem:

> [PROBLEM 5–15]
> Suppose one wanted to select a three-member subcommittee from the ten candidates. Are there also 720 ways to do this?

The answer is no because order does not matter in this case. For example, let's say that three of the people to be chosen are named Chris, Lucy, and Rachel. There are 3! ways (3! = 3 × 2 × 1 = 6) to select these three specific people to fill the three positions on the subcommittee:

Member (1) ↓	Member (2) ↓	Member (3) ↓
Chris	Lucy	Rachel
Chris	Rachel	Lucy
Lucy	Chris	Rachel
Lucy	Rachel	Chris
Rachel	Chris	Lucy
Rachel	Lucy	Chris

When making a subcommittee, however, the order of the selections is irrelevant. It only matters that three are chosen. It is irrelevant, for instance, whether the order is Chris, Lucy, Rachel or Rachel, Lucy, Chris. The people on the subcommittee are the same ones. The above six permutations, therefore, can be reduced to a combination of three distinct people. This is why this type of arrangement is, in fact, called a combination, rather than a permutation. It can be defined as an arrangement that is put together with no regard to order. In the case of a three-member subcommittee, the redundant permutations—3!—must be eliminated from the 720 possible selections that can be made. How many of these are there? To determine this, we simply cancel out the redundant permutations from 720 through division:

$$\frac{720}{3!} = \frac{720}{6} = 120$$

We point out that the denominator in this case is r!—the number of positions to be filled. We note as well that the 720 in the numerator was produced by our previous formula—namely, n!/(n-r)!. So, the general formula for a combination is:

$$\frac{n!}{(n-r)!\,r!}$$

Clearly, permutations and combinations form a binary opposition, revealing a structural difference between an ordered-versus-unordered arrangement. In this case the ordered arrangement is the unmarked form and, hence, the reason why permutations are taught first. Follow-up problems, with exploratory features, can now be easily composed to reinforce knowledge of combinatory structure and the applications of relevant formulas. Here are two examples, which are self-explanatory:

[PROBLEM 5–16]

In how many ways can you arrange the following:

(1) six friends in a row at a movie theater?

(2) the same six friends if one of them must have an aisle seat?

(3) the six friends in the first three seats of the row?

> [PROBLEM 5–17]
> A team of five hockey players must be chosen from 25 people. How can this be done?

Problems in Probability

To teach the structural principles involved in representing probability problems it is best to start with concrete real-world examples (that is, through contextualization). If one tosses a coin, it will, of course, fall as either a head or a tail. If the outcomes are equally likely—that is, if the coin has not been modified in any way—then the probability of each one occurring is 1 in 2, or in fraction form, 1/2. This fraction says that out of two possible outcomes (the denominator) the chance of getting a head is 1 in 2 as is the chance of getting a tail. Similarly, if we have a die, which has six numbers on it, the probability of throwing a 1 is 1/6, of throwing a 2, also 1/6, and so on. The branch of mathematics that deals with this kind of event is called *probability theory*. Probability is based on the study of permutations and combinations and is the foundation for statistics.

The probability of an outcome is represented by a number between 0 and 1 in the form of a fraction or corresponding decimal: 1/2 = 0.5, 1/6 = 0.16, etc. The outcome 0 indicates certainty that an event will not occur and 1 indicates certainty that it will occur. Thus, if an event has n equally likely outcomes, and f of them are termed favorable, the probability of a favorable outcome is represented as f/n. More involved problems concern events in which the various possible outcomes are not equally likely. Some events may have infinitely many outcomes.

Let's look at a typical problem in probability involving equally likely outcomes:

> [PROBLEM 5–18]
> If a committee of 11 people consists of 2 adults, 5 teenagers, and 4 seniors, and one of them is to be chosen to be chair, what is the likelihood of that person being an adult?

There are two adults; thus: f = 2. The total number of people is 11. So, the probability of an adult being chosen as chair is: f/n = 2/11 = 0.18. Probability problems are solved in this way. They might get more complex, but the underlying principle to solve them remains the same. Incidentally, it is interesting to point out the new function that fractions were discovered to have by the inventors of probability theory—namely, that they are much more than simple models of partitions (as discussed in chapters 1 and 2); they are also models of events and their occurrences. Semiotically, this suggests that the two referential domains might have something in common.

Without going into any philosophical discussion here, suffice it to say that the use of the same signifier form (the fraction) to represent part-whole relations and probability events connects them ontologically.

In general, the probability that an event E will occur, when all the events are equally likely, is indicated with the symbol P(E). Now, what happens if there are two events that can occur, E and F? More precisely: what is the probability, P(E or F), that at least one of them will occur? Here's a typical problem based on this type of probability:

> [PROBLEM 5–19]
> What is its probability of drawing a nine or a heart from a regular deck of 52 cards?

We let E stands for the event that a heart will be chosen and F that a nine will be selected. What is the probability of choosing a heart? Including the jack, queen, and king, there are 13 hearts among the 52 cards. Therefore:

$$P(E) = 13/52$$

Now, what is the probability of choosing a nine? There are four suits and in each there is a nine. So, there are 4 nines among the 52 cards. Therefore:

$$P(F) = 4/52$$

The chances of choosing either is the sum of these two probabilities:

$$P(E) + P(F) = 13/52 + 4/52 = 17/52 = 0.327$$

In general, the probability of getting P(E or F) is the sum of the individual probabilities of each event:

$$P(E \text{ or } F) = P(E) + P(F)$$

Determining the probability of something may involve figuring out some combinatory outcome. Indeed, probability theory and combinatorics are interrelated. For example:

[PROBLEM 5–20]

Calculate the probability of rolling two dice that will produce a four.

First, we must look at the possible combinations of the two dice. Each die has the numbers from one to six on its faces. So, there are 6 possibilities for the first die and for each one of these there are 6 possibilities for the second die. In total, there are $6 \times 6 = 36$ outcomes. All of these can be easily charted:

First		Second		Outcome
1	+	1	=	2
1	+	2	=	3
1	+	3	=	4
1	+	4	=	5
1	+	5	=	6
1	+	6	=	7
2	+	1	=	3
2	+	2	=	4
etc.		etc.		etc.

The number 36 is our n in the formula f/n. Now, to get the value of f we must figure out how many of the 36 combinations produce the desired outcome of 4. The same question can be rephrased as follows: What two digits, from 1 to 6, when added together equal 4? Clearly, the most the first digit can be is 3, because if it is 4 then it cannot be combined with the second one. So, let's chart the possible combinations:

First		Second		Desired outcome
1	+	3	=	4
2	+	2	=	4
3	+	1	=	4

These three are the only possible combinations that produce a four. So, the probability of rolling a four with two dice is: $f/n = 3/36 = 1/12 = 0.083$.

Let's look at one last problem of this type:

[PROBLEM 5–21]

Four cards labeled 1, 2, 3, and 4 are placed in a box. Four other cards, also labeled 1, 2, 3, and 4, are put in another box. If you are allowed to draw one card from each box, blindfolded, what is the probability of drawing two cards that add up to 4?

Again, we have to compute the number of ways a total of 4 can be generated by pairing two cards. The pairings are specified below:

First Box	Second Box	Outcome	Outcome = 4?
1	1	2	no
1	2	3	no
1	3	4	yes
1	4	5	no
2	1	3	no
2	2	4	yes
2	3	5	no
2	4	6	no
3	1	4	yes
3	2	5	no
3	3	6	no
3	4	7	no
4	1	5	no
4	2	6	no
4	3	7	no
4	4	8	no

The total number of pairings (events) is $4 \times 4 = 16$ (that is, for each card drawn from the first box there are four cards that can be matched with it). Of these, as can be seen, there are three pairings that add up to 4. Thus, the probability of drawing a pair that adds up to 4 is: $3/16 = 0.18$.

Now, as a related problem: what is the probability of drawing a 3 or 4? The total number of draws is, again, 16. But, the chart now shows that there are 5 possible outcomes:

First Box	Second Box	Outcome	Outcome = 3 or 4?
1	1	2	no
1	2	3	yes
1	3	4	yes
1	4	5	no
2	1	3	yes
2	2	4	yes
2	3	5	no
2	4	6	no
3	1	4	yes
3	2	5	no
3	3	6	no
3	4	7	no
4	1	5	no
4	2	6	no
4	3	7	no
4	4	8	no

Thus, the probability is 5/16 or 0.31 or approximately, one in three draws. Incidentally, if we compute the probabilities separately and then add them up we would get the same result. The probability of drawing a 4 is 3/16. The probability of drawing a 3 is 2/16. If we add these together we get: $3/16 + 2/16 = 5/16 = 0.31$.

Incidentally, it was gambling that let to the development of probability theory. In the mid-seventeenth century, the Chevalier de Méré, a high-living nobleman with a compulsion for gambling, is said to have asked his good friend, the mathematician Blaise Pascal (1623–1662), to counsel him on how to play advantageously. The reasoning used by Pascal can be reformulated as follows. On a roll of two dice, there are, of course, 36 possible pairings:

Die 2

	+	1	2	3	4	5	6
	1	2	3	4	5	6	7
	2	3	4	5	6	7	8
Die 1	3	4	5	6	7	8	9
	4	5	6	7	8	9	10
	5	6	7	8	9	10	11
	6	7	8	9	10	11	12

The Chevalier's favorite game strategy was to bet on 12 (produced only by a double 6 as the chart shows) on the basis of 24 tosses of two dice. Pascal told the Chevalier that since there is only one possible combination that produces a 12 and 35 that do not, the probability of getting one of the other numbers was 35/36. Now, since the Chevalier used 24 tosses as his game plan, Pascal indicated that the 35/36 probability would occur 24 times. Therefore, the Chevalier should expect to lose often.

Appropriate exploratory problems that can easily be associated with this historical incident can easily be composed:

> [PROBLEM 5–22]
> What is the probability of rolling a five with two dice?

> [PROBLEM 5–23]
> What is the probability of rolling either a five or a two with the two dice?

Heuristic Problem-Solving

As we have seen in this chapter, one of the most important forms of pedagogical PS is the heuristic one, a form that allows students to investigate mathematical structure through exploration. Heuristic PS also provides insights into the nature of conceptualization and the various "logics" behind it. The Greek philosophers and mathematicians had claimed that many concepts are formed in one of two fundamental ways—by induction or by deduction. The former involves reaching a general conclusion from observing a recurring pattern; the latter involves reasoning about the consistency or concurrence of a recurring pattern. Aware that other types of concepts existed (such as those found in poetry, the arts, music, etc.), they argued that induction and deduction were particularly apt in explaining logical and mathematical thinking. An example of the use of inductive reasoning is problem [5–4] above, which asked us to derive a formula for determining the number of degrees in the angles of a polygon. We arrived at the formula by observing individual cases of polygons and then extrapolating the pattern inherent in them. This "generalization-by-extrapolation" process is the sum and substance of inductive thinking. An example of the use of deductive reasoning is problem [5–1]. With this problem we were able to prove that the vertically-opposite angles formed when two straight lines intersected are equal, no matter what their size. We did this simply by representing the angles with variables (symbols standing for *any* number). This "generalization-by-demonstration" process is the sum and substance of deductive thinking.

Induction and deduction do, clearly, play a significant role in the formation of mathematical concepts. But they hardly explain how the entire range of such concepts comes about, as discussed in chapter 3. As Peirce argued (discussed again in chapter 3) informed hunches and guesses play a much more central role in the origination of mathematical ideas than do induction and deduction. These constitute a type of open-ended logical reasoning that he called abduction. We have used a number of problems in this book requiring abductive reasoning. It is nevertheless instructive to go over what is involved again here. Consider the following classic puzzle—a puzzle that gave origin to the expression "thinking outside the box:"

> [PROBLEM 5–24]
> Without letting your pencil leave the paper, draw four straight lines through the following nine dots:

Many students will approach this puzzle at first by attempting to join up the dots as if they were located on the perimeter (boundary) of a square or flattened box. But this reading of the puzzle does not yield a solution, no matter how many times one tries to draw four straight lines without lifting the pencil. A dot is always left over. It is at this point where abductive thinking comes into play: "What would happen if I extend one or more of the four lines beyond the box?" That hunch turns out, in fact, to be the relevant insight. One possible solution is as follows:

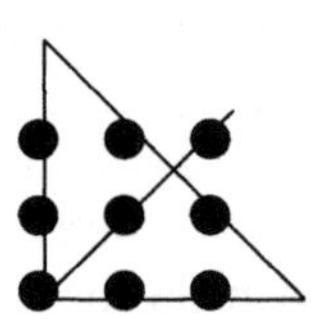

It should be obvious why this puzzle gave rise to the expression "thinking outside the box," which entered the English language around the middle part of the twentieth century when people in business and education started referring to it as a prototypical example of what creative or "lateral" thinking is all about. It continues to be cited by psychologists as an example of how the mind tends to impose unnecessary limitations upon methods of attacking problems. Who invented the puzzle? I have looked into several sources and have been able to trace it as far back as 1914, in the first edition of American puzzlist Sam Loyd's (1841–1911) *Cyclopedia of Puzzles*. But the principle it embodies is probably older, as Martin Gardner indicates in his 1960 edition of Loyd's *The Mathematical Puzzles of Sam Loyd*.

Now, once the solution principle behind the Nine-Dot problem has been uncovered, then other logical processes (inductive and deductive) come into play to help us generalize any intrinsic pattern hidden within the solution. The Nine-Dot puzzle is a 3 × 3 version of what can be called generally a Dot-Joining puzzle. We can now compose two relevant follow-up problems for heuristic purposes:

[PROBLEM 5–25]

Can you solve a Sixteen-Dot (4 × 4) puzzle version?

[PROBLEM 5–26]

Can you solve a Twenty-Five-Dot (5 × 5) puzzle version?

The solutions reveal that six lines are needed to solve the 4 × 4 version and eight the 5 × 5 version (as readers can confirm for themselves). Can we generalize further? By making the Nine-Dot puzzle as complex as we desire (increasing the number of dots to 16, 25, 36, 49, etc.), a pattern seems to emerge through inductive inspection. This pattern can be charted as follows:

Dots	*Lines Required*
3 × 3	(3 + 1) = 4
4 × 4	(4 + 2) = 6
5 × 5	(5 + 3) = 8
6 × 6	(6 + 4) = 10
...	...
n × n	n + (n - 2) = 2n - 2

I should point out that I have not tested this pattern beyond a 6 × 6 version of the puzzle. As with all inductively-derived formulas, there is no way to be sure that the formula works all the time. It thus can be called, simply, a working formula.

In many ways, teaching PS in mathematics is akin to teaching a new language. That language may have as its referential domain the properties related to triangles, to arrangements of things, to the probability of something taking place, and the like; but it is still a language. As the American linguist Charles F. Hockett (1967: 6) aptly pointed out, this may not always be recognized by mathematicians: "This ignorance on the part of the average mathematician has no bearing on the quality of his mathematics; but ultimately the language-like nature of mathematics is of basic importance, since it is the most critical clue to an understanding of the place of mathematics in the physical universe of which mathematics, mathematicians, language, linguists and all of us are a part." The "semiotic mind," so to speak, investigates how mathematical ideas are interrelated—one suggesting another. In effect, all math is experimentation with generalization. The end result is discoveries and applications that could not have been possible otherwise. Consider the number represented by the alphabet character e. This is defined as the limit of the expression $(1 + 1/n)^n$ as n becomes large without bound. Its limiting value is approximately 2.7182818285. Now, one may ask, what possible connection does this number have with other areas of mathematics? As it turns out, it forms the base of natural logarithms; it appears in equations describing growth and change; it surfaces in formulas for curves; it crops up frequently in probability theory; and it appears in formulas for calculating compound interest (Maor 1994).

6
SEMIOTIC PEDAGOGY

As far as the laws of mathematics refer to reality, they are not certain, and as far as they are certain, they do not refer to reality.
Albert Einstein (1879–1955)

Mathematics is all about studying our innate number sense. It started out rather simply as the art of modeling how we count instinctively, transforming natural counting signs (fingers, sticks, etc.) into symbolic ones (numerals). It eventually became the discipline examining the properties of these very signs. Because of the importance of counting to human life, across time and across the world mathematics has always been considered to be a central part of one's upbringing and education. And across cultures, PS has always constituted the primary means for imparting knowledge of mathematical principles. The assumption implicit in this practice has always been that we come to learn mathematics by exploring its structure through problems. Indeed, one of the oldest "math textbooks" is a 1650 BCE Egyptian Papyrus (mentioned in the preface to this book) containing a collection of math problems probably intended for Egyptian youth. Eighteen and a half feet long and thirteen inches wide, the manuscript is referred to either as the *Ahmes Papyrus*, after the Egyptian scribe who copied it, or the *Rhind Papyrus*, after Scottish lawyer and antiquarian, A. Henry Rhind (1833–1863), who purchased it in 1858 while vacationing in Egypt. Since the manuscript is a copy, it is estimated that it may actually have been written nearly four thousand years ago, during the same period in which another famous record of Egyptian mathematics, the *Moscow Papyrus* (named after its current location), was written. In addition to eighty-four challenging mathematical problems, the Papyrus contains tables for the calculation of areas and the conversion of fractions, elementary sequences, linear equations, and extensive information about measurement. The earliest known symbols for addition, subtraction, and equality are also found in this truly remarkable work (Gillings 1972: 246–247).

If it is true that the Papyrus was a math textbook, as many claim, then it shows how crucial PS was considered to be in the ancient world in the teaching of math. The Papyrus contains eighty-four problems in arithmetic, geometry, and algebra, together with their solutions. Its main pedagogical objective seems to have been to impress upon Egyptian youth, by pedagogical illustration, that the exercise of accurate reckoning was the key to successful problem-solving (Olivastro 1993: 31–64, Gillings 1961, 1962, 1972). As I have argued elsewhere, the work also brings out the ontological interconnection between PS and discovery (Danesi 2002). Solving Problem 56, for instance, leads to an insight that was original for the era in which the papyrus was written—namely, that the height of a pyramid can be related to the size and slope of each of its triangular walls.

PS has been the heart and soul of mathematics since the beginning of time. And, as mentioned in the preface to this book, there is no evidence of so-called math phobia in the ancient world, or in any other era for that matter, other than the present one. The view espoused in this book has been that to recapture the instinctual love for PS, implicit in the Rhind Papyrus and similar ancient texts, one has to deconstruct it into its basic sign-based components (as did the Papyrus, by the way). I have attempted to make the argument that this will make math education more effective because it provides insights into how learning and representational processes overlap and intertwine. For example, I have argued throughout that novel learning tasks are more apt to be more comprehensible if they are presented iconically (through concretization, contextualization, etc.) because the most basic signs are iconic ones. That is the central notion in semiotic pedagogy.

In this final chapter I will synthesize the main ideas espoused in previous chapters, primarily by illustration. I will also visit one more time a subtext that I have simultaneously attempted to interweave throughout—namely, that mathematics is ipso facto a semiotic modeling activity, whether mathematicians have known this or not.

Representing Structure

As discussed throughout this book, the crux to successful PS involves representing some structural feature or pattern implicit in a problem that will bring out its characteristics for inspection and analysis. As another example of this basic principle of semiotic pedagogy, consider the following computations:

$2 \times 9 = 18$	and	$1 + 8 = 9$
$3 \times 9 = 27$	and	$2 + 7 = 9$
$4 \times 9 = 36$	and	$3 + 9 = 9$
$5 \times 9 = 45$	and	$4 + 9 = 9$
…		
$12 \times 9 = 108$	and	$1 + 0 + 8 = 9$
$123 \times 9 = 1107$	and	$1 + 1 + 0 + 7 = 9$
$1{,}245 \times 9 = 11{,}205$	and	$1 + 1 + 2 + 0 + 5 = 9$
$12{,}459 \times 9 = 112{,}131$	and	$1 + 1 + 2 + 1 + 3 + 1 = 9$
…		

These computations show that the digits of any multiple of 9, when added together, produce 9 (or a multiple of 9). Once detected, the pattern is rendered communicable through a sign, representamen, or annotation such as the statement made in the previous sentence. And, as we have seen throughout this book, can thus be used to compose follow-up problems for pedagogical purposes. Here is an illustrative one:

[PROBLEM 6–1]
Determine if the numbers 477, 648, and 8,766 are multiples of 9.

In a fundamental sense, mathematics is the science of detecting and encoding patterns. As Pythagoras claimed, this science is a magical one, for the reason that it seems to reveal the secrets of the universe. Some of that sense of mystery and magic continues on within us today when we solve some difficult problem or finally get the point of some formula. The reasoning process involved tends to produce what psychologists call an "Aha" effect—a feeling of satisfaction and wonderment at once. This very feeling was, more than likely, the main reason why mathematics was considered more of a craft, akin to alchemy and astrology, than a rigid science in the ancient world.

Dwelling on this point a little further, it is instructive to consider a classic example of the interplay that exists between representation and discovery in math from the seventeenth century, when the French mathematician Blaise Pascal decided to examine the numbers obtained in the expansion of $(a + b)^n$ where n = {0, 1, 2, 3, 4, 5, ...}. He found a hidden pattern within the expansions—a pattern that led to various discoveries in number theory:

$$
\begin{aligned}
(a+b)^0 &= 1\\
(a+b)^1 &= a+b\\
(a+b)^2 &= a^2+2ab+b^2\\
(a+b)^3 &= a^3+3a^2b+3ab^2+b^3\\
(a+b)^4 &= a^4+4a^3b+6a^2b^2+4ab^3+b^4\\
(a+b)^5 &= a^5+5a^4b+10a^3b^2+10a^2b^3+5ab^4+b^5\\
(a+b)^6 &= a^6+6a^5b+15a^4b^2+20a^3b^3+15a^2b^4+6ab^5+b^6\\
(a+b)^7 &= a^7+7a^6b+21a^5b^2+35a^4b^3+35a^3b^4+21a^2b^5+7ab^6+b^7\\
(a+b)^8 &= a^8+8a^7b+28a^6b^2+56a^5b^3+70a^4b^4+56a^3b^5+28a^2b^6+8ab^7+b^8
\end{aligned}
$$

Since there are two terms, a and b, it is called a *binomial expansion*. If we look at the numerical coefficients in the expansion, the shape of a triangle in outline form seems to jut out. That iconic suggestion is what probably led

Pascal to make his discovery. In the top row, the numerical coefficient is 1; in the row below the numerical coefficients of a and b are both 1; in the row below that the numerical coefficients are 1, 2, and 1; and so on. These can be highlighted as follows.

$$(a + b)^0 = \boxed{1}$$
$$(a + b)^1 = \boxed{1}a + \boxed{1}b$$
$$(a + b)^2 = \boxed{1}a^2 + \boxed{2}ab + \boxed{1}b^2$$
$$(a + b)^3 = \boxed{1}a^3 + \boxed{3}a^2b + \boxed{3}ab^2 + \boxed{1}b^3$$
$$(a + b)^4 = \boxed{1}a^4 + \boxed{4}a^3b + \boxed{6}a^2b^2 + \boxed{4}ab^3 + \boxed{1}b^4$$
$$(a + b)^5 = \boxed{1}a^5 + \boxed{5}a^4b + \boxed{10}a^3b^2 + \boxed{10}a^2b^3 + \boxed{5}ab^4 + \boxed{1}b^5$$

etc.

Extracting these coefficients from the other terms in the expansion, leaving them in their respective rows, and deleting the literal coefficients produces the following triangular arrangement:

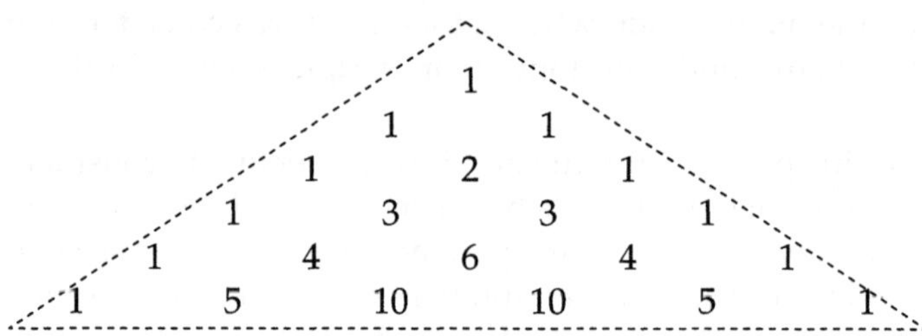

The diagram is, of course, only a slice of the triangle, which has infinite dimensions. Now it is this arrangement that likely led Pascal to discover a hidden structural feature—each row begins and ends in 1 and the numbers in a row are the sum of the two numbers above it (for example, each of the two 3's in the fourth row of our slice is the sum of the two right above it. Take, for instance, the left one:

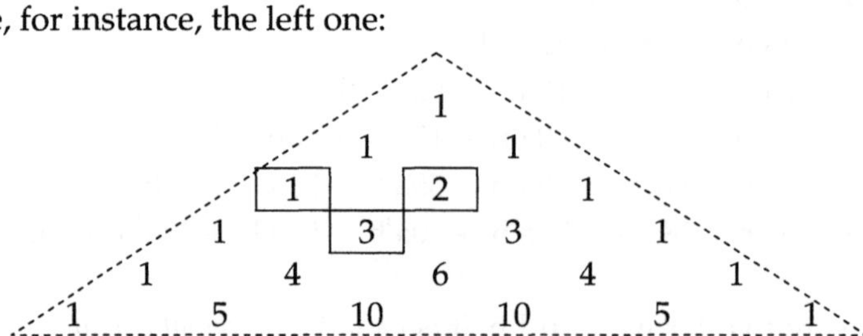

Similarly, each of the two 10's in the bottom row is the sum the two numbers just above it. Take, for instance, the right one:

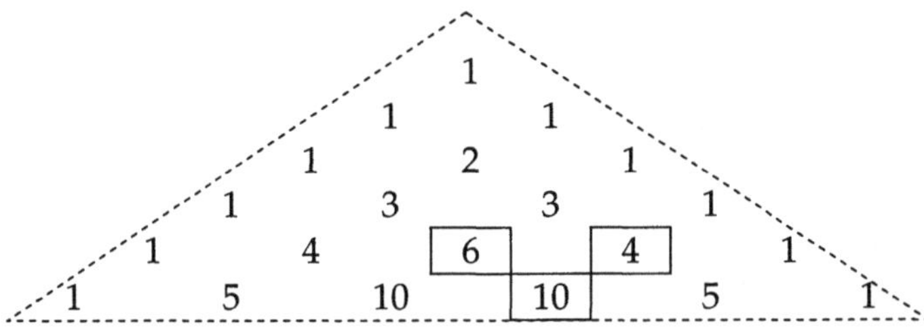

That is the "hidden pattern" and it is that very pattern that makes it an infinite triangle, since we could continue to generate numbers within it with this simple rule ad infinitum: *Any number in a row is produced by the sum of the two numbers above it.* The interesting thing is that Pascal fleshed out this pattern by taking elements from the binomial expansion and arranging them in a representamen that revealed the hidden structure. That is the sum and substance of mathematics and it is, needless to say, the sum and substance of semiotic analysis. Incidentally, when Isaac Newton was a student in his early twenties, he noted that the exponent of the binomial expression indicated the row of the triangle. For example, in the expansion of $(a + b)^4$, which is $a^4 + 4a^3b + 6a^2b^2 + 4ab^3 + b^4$, the numerical coefficients (1, 4, 6, 4, 1) coincide with the fourth row of numbers in the triangle; the coefficients in the expansion of $(a + b)^5$ coincide with the fifth row of numbers in the triangle; those in the expansion of $(a + b)^6$ with the sixth row; and so on.

The triangle is known as *Pascal's triangle,* logically enough. For the sake of historical accuracy, it should be mentioned that the pattern it displays was known as early as 1303, since it is described in the *Precious Mirror of the Four Elements,* published in that year by the Chinese mathematician Chu Shih-chieh. There is no evidence, however, to suggest that Pascal was aware of this text. The two detections of the pattern seem to have occurred autonomously. Now, many discoveries have been made possible by this triangle—discoveries that would have been unimaginable without it. It is noticeable, for instance, in the calculation of probabilities. If a coin is to be tossed eight times in a row, there are two possibilities for the coins to be the same, all heads or all tails (H = head, T = tails):

(1) H H H H H H H H ← *Eight heads thrown in a row*
(2) T T T T T T T T ← *Eight tails thrown in a row*

Another way to describe these two outcomes is to say that the former consists of "no tails" and the latter of "no heads." Now, let's go on to the next possibility, namely that there are eight outcomes composed of seven heads and one tail. These can be shown as follows:

(1) H H H H H H H T
(2) H H H H H H T H
(3) H H H H H T H H
(4) H H H H T H H H
(5) H H H T H H H H
(6) H H T H H H H H
(7) H T H H H H H H
(8) T H H H H H H H

For six heads and two tails, there are 28 outcomes (as readers may wish to verify for themselves); for five heads and three tails, there are 56 outcomes; and so on. Incredibly, these outcomes coincide with the numbers in the eighth row of Pascal's triangle (1, 8, 28, 56, 70, 56, 28, 8, 1): that is, there is 1 outcome consisting of no tails; 8 outcomes of one tail; 28 outcomes of two tails; and so on. Altogether, the total number of possible outcomes of tails is:

$$1 + 8 + 28 + 56 + 70 + 56 + 28 + 8 + 1 = 256$$

So, the probability of getting all heads and no tails in eight tosses is 1/256; the probability of seven heads and one tail is 8/256 = 1/32; the probability of six heads and two tails is 28/256 = 7/64; and so on. What is even more incredible is the discovery that, in general, the outcomes of n tosses coincide with the numbers in the nth row of Pascal's triangle: the outcomes of eight tosses coincide with the numbers in the eighth row of the triangle; the outcomes of nine tosses coincide with the numbers in the ninth row of the triangle; and so on.

The discovery of mathematical pattern raises some fundamental questions, already broached in previous chapters. Are we constituted by our nature to discover it? Are the patterns discovered by mathematicians mirrors of the patterns hidden in Nature? One thing is certain—the discovery of pattern (when it happens) is a fascinating event in and of itself. And it is the sense of magic that solving problems is meant to elicit through the use of appropriate representations of the problems. The latter are like little portraits of problem structure; and like real portraits, they bring out features that may otherwise not have been evident in their subjects. A diagrammatic model of a real-world situation or object is always a magical construct, so to speak, because it fleshes out features that are not always detectable by direct perception. Since models can be changed or manipulated, we can see how the real version might be affected by modifications or varying conditions.

The Interconnectedness Principle

The discovery of related patterns is likely the result of what semioticians call sign-interconnectedness, namely that representations of objects in one domain contain in them structural information about objects in other domains. It is worthwhile to take a brief digression on the power of this principle to explain related phenomena, such as PS and mathematical discovery, since it is ultimately what semiotic pedagogy attempts to bring to awareness in learners.

The interconnectedness principle posits that expressions, symbols, representations, and traditions are interconnected to each other through *semiosis*—the instinctive tendency to construct signs to stand for reality or, more accurately, for human perceptions of that reality. Science is an attempt to render visible with our inner eye those things we can never see with our real eye—atoms, sound waves, gravitational forces, magnetic fields, etc. The trace to this "inner vision" is semiosis. And the representation tends to be metaphorically-guided (chapter 4). This is why sound waves are said to *undulate* through empty space like water waves ripple through a still pond; atoms to *leap* from one quantum state to another; electrons to *travel in circles* around an atomic nucleus; and so on. The physicist K. C. Cole (1984: 156) puts it as follows:

> The words we use are metaphors; they are models fashioned from familiar ingredients and nurtured with the help of fertile imaginations. "When a physicist says an electron is like a particle", writes physics professor Douglas Giancoli, "he is making a metaphorical comparison like the poet who says "love is like a rose". In both images a concrete object, a rose or a particle, is used to illuminate an abstract idea, love or electron.

As Robert Jones (1982: 4) has also pointed out, for the scientist metaphor serves as "an evocation of the inner connection among things." It is relevant to note that the philosopher of science, Fernand Hallyn (1990), has identified the goal of science as that of giving the world a "poetic structure."

As a concrete example of how the scientist uses an innate sense of semiosis, or semiotic logic, to represent physical reality, consider the origin of atomic theory. It was Ernest Rutherford (1871–1937) who speculated, in effect, that atomic structure mirrored the structure of the solar system—a theory that was consistent with the ancient Greek concept of the cosmos as having the same structure at all its levels, from the microcosmic (the atom) to the macrocosmic (the universe). Without going here into the history of atomic theory, suffice it to say that it is mind-boggling to think that such a simple conceptual linkage has led to real knowledge about the atom, both through falsification of elements of the theory and through expansion of other parts. It appears that the semiotic brain is a discovery organ—an organ designed to seek hidden pattern through connections based on sense inferences and hunches. The astounding "predictive power" of mathematics lies, arguably, in the fact that the very representation of pattern, such as that hidden in the binomial expansion, reifies hidden connections among real things. Perhaps, as the mathematician's representamina suggest, the human brain is designed to discover patterns in the world serendipitously because it already has such patterns built into it.

The founder of mathematics as a science of pattern, Pythagoras, certainly was aware of the interconnectedness principle. As is well known Pythagoras founded a society, known as the Pythagorean Brotherhood, around 500 BCE to study this principle, believing firmly that knowledge of the universe could only come from contemplating how it revealed itself through mathematical symbols. The term "Pythagorean Brotherhood" is in all likelihood a mistranslation, because Pythagoras encouraged women to participate fully in his so-called "brotherhood." Late in life, he married one of his students, Theano. An accomplished cosmologist and healer, Theano headed the Pythagorean society after her husband's death, and even though she faced persecution, continued to spread the Pythagorean philosophy throughout Egypt and Greece alongside her daughters.

One of the patterns studied by the Pythagoreans was that of *figurate numbers*—numbers that can be displayed geometrically. For example, square integers, such as 1^2 (= 1), 2^2 (= 4), 3^2 (= 9), and 4^2 (= 16) can be displayed with square arrangements of objects:

1 4 9 16

As a consequence of displaying square numbers in this way, the Pythagoreans discovered that square numbers are equal to the sum of consecutive odd integers:

1	=	1
4	=	1 + 3
9	=	1 + 3 + 5
16	=	1 + 3 + 5 + 7
25	=	1 + 3 + 5 + 7 + 9

And so on

Why is this so? It is so because to form each new square figure a successive odd number of elements to the preceding figure must be added—for example, to the square figure for the number 1 above 3 new little dark squares must added to produce the next figure for 4; then to the square for the number 4 above, 5 more little dark squares must be added to produce the next figure for 9; and so on. If we examine the square of 16 more closely, we can partition it off as follows, using this discovered property:

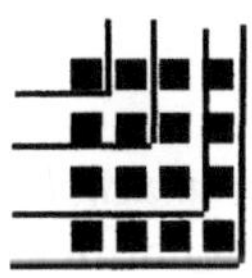

16

This display now shows clearly that 16 is the sum of the first four odd numbers, since it is made up successively of 1 + 3 + 5 + 7 additions of little dark squares. By the way, the Pythagoreans called the odd numbers *gnomic* because the partitioning method looks like a carpenter's square known as a *gnomon*.

The Pythagoreans were intrigued by such findings and, consequently, felt impelled to study all kinds of figurate numbers. The triangular ones held a special kind of place in their experiments with the interconnectedness between numeration and geometry. Here are the first four triangular numbers:

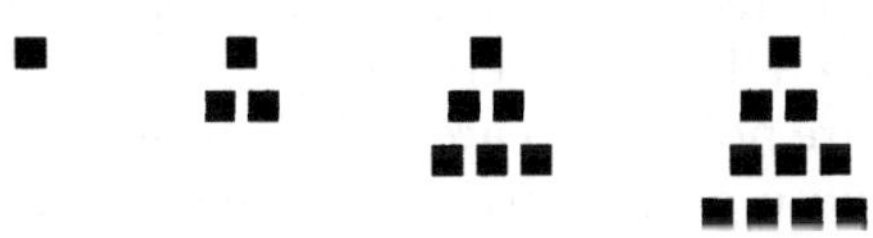

1 3 6 10

The number of little dark squares in the figure for the first triangular number 1 is 1; the number of such squares that make up the figure for the second triangular number 3 is 1 + 2; the number of such squares that make up the third triangular number 6 is 1 + 2 + 3; and so on. Why is this so? It is so because each successive triangular number figure above is obtained by adding a row of little dark squares containing one more square than the number of squares in the bottom row of the previous triangular number. This leads to the discovery that the nth triangular number is the sum of the first n counting numbers:

$$
\begin{aligned}
1 &= 1 \\
3 &= 1+2 \\
6 &= 1+2+3 \\
10 &= 1+2+3+4 \\
15 &= 1+2+3+4+5 \\
21 &= 1+2+3+4+5+6 \\
&\dots \\
n^{th} &= 1+2+3+4+5+6+\dots
\end{aligned}
$$

The Pythagoreans also came up with the notions of *amicable* and *perfect numbers*. The numbers 284 and 220, for instance, are called amicable because the proper divisors of one of them when added together produce the other. The proper divisors of 284 are 1, 2, 4, 71, and 142 (as students may be asked to check for themselves), and the proper divisors of 220 are 1, 2, 4, 5, 10, 11, 20, 22, 44, 55, and 110 (again, as students may wish to check for themselves). Now, we notice that:

$$220 = 1 + 2 + 4 + 71 + 142$$
$$284 = 1 + 2 + 4 + 5 + 10 + 11 + 20 + 22 + 44 + 55 + 110$$

A perfect number, on the other hand, is one that equals the sum of its proper divisors, with the exception of itself. For example, the proper divisors of the number 6 are 1, 2, and 3. Now, if we add these together we get the number: $6 = 1 + 2 + 3$. The next perfect number is 28. Its proper divisors are 1, 2, 4, 7, and 14, and $1 + 2 + 4 + 7 + 14 = 28$. Very few perfect numbers have been discovered. Numbers that are not perfect are called *excessive* or *defective*. An excessive number is one whose proper divisors, when added together, produce a result that exceeds its value. The number 12, for example, is excessive because the sum of its proper divisors, 1, 2, 3, 4, and 6 ($1 + 2 + 3 + 4 + 6 = 16$) exceeds its value. A defective number is one whose proper divisors, when added together, produce a result that is smaller than its value. One example is 8, since the sum of its proper divisors 1, 2, and 4 ($1 + 2 + 4 = 7$) is less than its value.

Incidentally, this "Pythagorean sense" for studying number patterns can be easily built into exploratory problems for classroom use. Here are a few illustrative ones. They are self-explanatory:

[PROBLEM 6–2]

Write down the squares of numbers from 1 through 20. What seems to be characteristic about the square of an odd number or of an even number? What can you predict about 22^2?

[PROBLEM 6–3]

Is it possible to write the numbers from 2 to 100 as the sums of squares? If so, do you detect any pattern? Here are a few sample cases:

(1) $2 = 1^2 + 1^2 = 1 + 1$
(2) $5 = 1^2 + 2^2 = 1 + 4$
(4) $10 = 1^2 + 3^2 = 1 + 9$

Bringing out the interconnectedness of a number with other numbers and with geometrical figures, not to mention with other areas of math, is the essence of semiotic pedagogy. Metamathematical awareness, as it has been defined here, consists in seeing the interconnectedness among sign systems within the different areas of math—arithmetic, algebra, etc. Such awareness is stimulated typically in learners by emphasizing that these relations can be deconstructed into sets of binary oppositions (plus versus minus, addition versus subtraction, etc.) or multiple-leveled oppositions (addition versus subtraction versus multiplication versus subtraction, etc.). The essence of semiotic pedagogy is, as we have seen, to organize the teaching syllabus according to such oppositions so as to bring out their relevance to the PS process.

Logical Strategies

The main purpose of this book has been to show how successful PS inheres in devising an appropriate representation of problem content, and then to argue that the representation itself accordingly identifies which logical or cognitive strategy (inductive, deductive, inferential, metaphorical, or otherwise) is to be used in solving the problem. The history of the great problems of mathematics has, actually, documented this very fact. Take, as an example, the logical strategy called *reductio ad absurdum*, a strategy devised originally by Zeno of Elea (c. 490–430 BCE) so that he could stump his debating opponents by deriving absurd conclusions from their opinions (hence, the term "reduction to the absurd"). But a version of this strategy devised by Euclid to prove the infinity of prime numbers went far beyond Zeno's initial (and likely mischievous) usage of this strategy, installing it into mathematics as a primary logical technique for solving problems and for exploring mathematical structure generally.

I have often used Euclid's proof in my classes to great advantage. To start off, I remind students that the integers are divided into those that can be decomposed into factors—composite numbers—and those that cannot. The numbers 12, 42, and 169, for instance, are all composite because they are the products of smaller factors: 12 = 2 × 2 × 3, 42 = 7 × 2 × 3, 169 = 13 × 13. The prime numbers are those that cannot be decomposed in this way. They are the factors in composite numbers. Now, even a cursory examination of the number line—{0, 1, 2, 3, 4, 5, 6, 7, 8, 9, 10, ...}—reveals that there are fewer and fewer primes on it as the numbers increase. Thus, it appears logical to conclude that the primes must come to an end at some point. "Common sense" would also have it that if a number is big enough, it must be the product of smaller prime factors. But, using the reductio ad absurdum strategy, Euclid proved that this is not so.

He started with the assumption that there may indeed be a finite set of primes, labeling them as follows (P = set of all primes).

$$P = \{p_1, p_2, p_3, \dots p_n\}$$

This representation is crucial to unlocking the logical strategy to be used in the proof. The symbol p_n stands for the last (largest) prime; each of the other symbols stand for a specific prime. So, $p_1 = 2$, , $p_2 = 3$, and so on. Concretely, the set would look like this:

$$P = \{2, 3, 5, 7, 11, \dots p_n\}$$

Euclid then had one of the illuminations that comes from considering pattern: What kind of number would result from multiplying all the primes in the set? Let's call this number N:

$$N = \{p_1 \times p_2 \times p_3 \times \dots \times p_n\}$$

Concretely, this would look like this:

$$N = \{2 \times 3 \times 5 \times \dots \times p_n\}$$

N would, of course, be a composite number because it can be factored into smaller prime factors—p_1, p_2, etc. Then, probably by contemplating his representation of N, Euclid had another flash of insight, Euclid added 1 to this product: $(p_1 \times p_2 \times p_3 \times \dots \times p_n) + 1$. Let's call the number produced in this way M:

$$M = \{p_1 \times p_2 \times p_3 \times \dots \times p_n\} + 1.$$

Clearly, M is not decomposable, because when any of the prime factors available to us $\{p_1, p_2, p_3, \ldots p_n\}$ are divided into it, a remainder of 1 would always be left over. So, the number M is either: (1) a prime number that is not in P and obviously much greater than p_n, or (2) a composite number with a prime factor that, as just argued, cannot be found in the set $\{p_1, p_2, p_3, \ldots p_n\}$ and thus also greater than p_n. Either way, there must always be a prime number greater than p_n. In this way, Euclid showed that the primes never end.

Euclid's proof puts on brilliant display the power and mysterious appeal of logical thinking. Incidentally, ever since its formulation, Euclid's proof has motivated myriads of mathematicians to come up with a formula for generating the prime numbers—but so far to no avail. Prime numbers are the "building blocks" of the whole architecture of arithmetic. So, they require some historical mention in any math class (as a contextualization technique). In this way, students can be taken into the heart of mathematics, so to speak (Whitlin and Whitlin 2002). For example, students can be introduced to the so-called Goldbach Conjecture. In a letter to the great mathematician Leonhard Euler in 1742, the mathematician Christian Goldbach (1690–1764) conjectured that every even integer greater than 2 could be written as a sum of two primes (Salem, Testard, and Salem 1992, 110–111):

$4 = 2 + 2$
$6 = 3 + 3$
$8 = 5 + 3$
$10 = 7 + 3$
$12 = 7 + 5$
$14 = 11 + 3$
$16 = 11 + 5$
$18 = 11 + 7$
...

No exception is known to Goldbach's Conjecture, as it has come to be known, but we have no valid proof of it either. Goldbach also conjectured that any number greater than 5 could be written as the sum of three primes:

$6 = 2 + 2 + 2$
$7 = 2 + 2 + 3$
$8 = 2 + 3 + 3$
$9 = 3 + 3 + 3$
$10 = 2 + 3 + 5$

11 = 3 + 3 + 5

...

Actually, one of the first to put forward an algorithm-device for identifying primes was the Greek mathematician Eratosthenes of Cyrene (c. 276–195 BCE). His device was an imaginary sieve by which a list of primes within the first 100 numbers can be identified by the process of elimination (as already discussed in chapter 2:

1	2	3	4	5	6	7	8	9	10
11	12	13	14	15	16	17	18	19	20
21	22	23	24	25	26	27	28	29	30
31	32	33	34	35	36	37	38	39	40
41	42	43	44	45	46	47	48	49	50
51	52	53	54	55	56	57	58	59	60
61	62	63	64	65	66	67	68	69	70
71	72	73	74	75	76	77	78	79	80
81	82	83	84	85	86	87	88	89	90
91	92	93	94	95	96	97	98	99	100

We start by crossing out every second number after 2. This eliminates all numbers that can be divided evenly by 2, except for 2 itself. Similarly, we cross out every third number after 3, since these are multiples of 3. This step eliminates all the numbers that can be divided evenly by 3, except for 3 itself. Then, since 4 is already crossed out, we move on to 5. We cross out every fifth number after 5. We continue this process. Numbers that are not crossed out can be thought of as having passed through a sieve (strainer) that has caught all the primes. So, the sieve ends up "trapping" 25 primes. Needless to say, it would be a gargantuan task to set up a sieve to identify the primes among the first 1000 numbers. The sieve anecdote is actually laden with pedagogical suggestions. One exploratory activity I assign after introducing the sieve concept is that of constructing a similar device (if any is possible) that might apply to detecting higher primes. The goal of the assignment is to get students to grasp the inner nature of one of the great unsolved problems of mathematics: Is there a formula or algorithm for generating primes, and only, primes?

The history of primes also presents opportunities for showing how PS and math discovery are interrelated. For example, the following problem is one I have used several times:

[PROBLEM 6–4]

Below are numbers that are all prime. They seem to show a pattern. What is the pattern? Does the pattern keep going?

31
331
3,331
33,331
333,331
3,333,331
33,333,331

The answer is that it does not keep going. The next number built with the pattern demonstrated by the others, 333,333,331, is actually a composite number: 17 × 19,607,843 = 333,333,331. A sub-purpose of the exercise is to show the student that sometimes induced pattern leads astray, especially if a formula for the pattern cannot be devised (as is the case here). It shows, in effect, that pure induction, as powerful as it often is, is actually limited as a logical strategy.

Another episode that I have often discussed in the classroom is the Mersenne formula, named for Marin Mersenne (1588–1648), a French priest and teacher. Mersenne proposed the formula $2^n - 1$, which, while it does not always generate a prime, does indeed generate a significant number of primes:

$n = 2$
$2^n - 1$
$2^2 - 1 = 4 - 1 = 3$ (prime)

$n = 3$
$2^n - 1$
$2^3 - 1 = 8 - 1 = 7$ (prime)

$n = 4$
$2^n - 1$
$2^4 - 1 = 16 - 1 = 15$ (composite)
...
$n = 19$
$2^n - 1$
$2^{19} - 1 = 524{,}288 - 1 = 524{,}287$ (prime)
...

The prime numbers generated by the formula are called, logically, Mersenne primes. The formula has peaked the interest of mathematicians since it was devised. After n = 30, the formula becomes unwieldy as a test for primality. Mersenne claimed that $2^{31} - 1$ was prime, but it was proved to be prime only much later in 1772; on the other hand, Mersenne also maintained that $2^{67} - 1$ and $2^{257} - 1$ were prime, but was proved later to be wrong. By 1914, mathematicians had discovered a prime number with the formula that was 29 digits long. Today, computers use the formula to keep searching for longer and longer primes. Called the Great Internet Mersenne Prime Search (GIMPS), and started by computer programmer George Woltman in 1996, volunteers from across the globe continue to search for larger and larger Mersenne primes with GIMPS (at least at the time of the writing of this book).

Euclid's proof of prime number infinity is a peculiar form of logic, qualitatively different from either induction or deduction. As mentioned several times in this book, the latter two have always been considered to be primary logical strategies in PS. Consider the following problem, which requires the use of inductive logic for its solution:

> [PROBLEM 6–5]
> What is the maximum number of pieces obtained when different cuts are made all the way across a circular cake (that is, without cutting through the same point more than twice)?

We don't need a cake, of course, to solve the problem. A circle will do, since it represents a cake in its essential outline. Let's carry out a number of cuts, or lines drawn across the circle, starting from 0 cuts and progressing to make one cut more each time. We can record the results in a table as we go along:

Cuts		Pieces	Increase
0	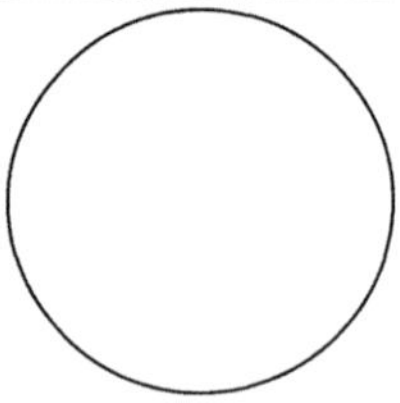	1	0

1	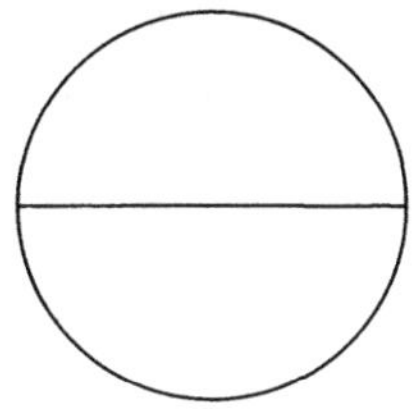	2	1
2	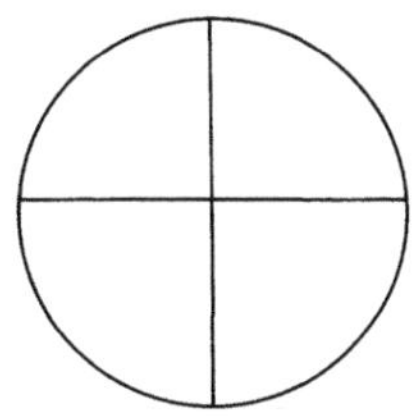	4	2
3	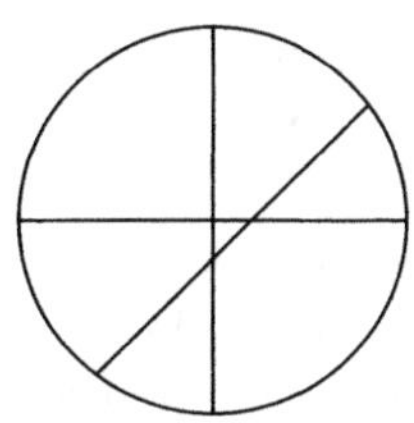	7	3

Examining the table carefully a pattern juts out before us—the number of pieces produced by each successive cut seems to increase in numerical order {0, 1, 2, 3,...}, paralleling the increase in the number of cuts. Will this pattern continue? Let's see whether it does or does not by trying out one more cut—the fourth one—and recording the outcome as before:

Cuts		Pieces	Increase
4	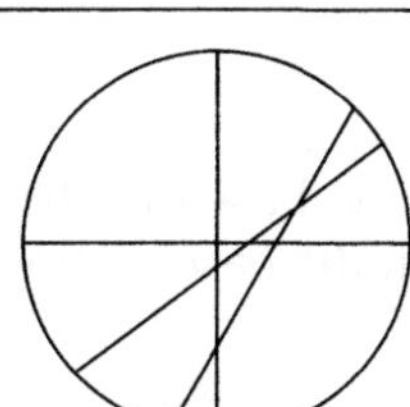	11	4

The pattern does indeed seem to continue. A fourth cut produces an increase in four pieces from the previous cut, from 7 pieces to 11. Thus, a fifth cut should produce an increase in five pieces from the previous fourth cut, that is, from 11 to 16. Students can be asked to confirm for themselves that this is indeed the case. We can now summarize the solution as follows:

Cuts	Pieces Produced	Increase from Previous Cut
0	1	0
1	2	1
2	4	2
3	7	3
4	11	4
5	16	5
...	...	...

It is obviously necessary to repeat this inductive experiment many times to see if the same pattern is always obtained. Since it is impractical to work with all possible cases, we can never be one hundred percent sure that our conclusion will apply to one and every case.

To arrive at the conclusion that something holds for every single case, a different type of proof is required, known as deductive. As we have seen several times in this book, this involves starting with several ideas that we either assume to be true or have previously established as true, and then using these to reach a foolproof conclusion by examining the general case. Suppose we assume every student in the ninth grade takes mathematics. If we know that Mary Jones is in the ninth grade, we can deduce that Mary Jones is taking mathematics. This specific conclusion is dependent upon the assumption with which we started our reasoning—namely that every student in the grade takes mathematics.

Consider the following problem in geometry, as a perfect example of deductive reasoning:

> [PROBLEM 6–6]
> Prove that the angle opposite the base of a triangle inscribed in a semicircle (half circle), with the base resting on the diameter, is equal to 90^0.

Notice that the problem does not ask us to experiment with different cases of triangles inscribed in semicircles. Rather it asks us to show that the condition it stipulates holds true for every single triangle inscribed in a semicircle. How do we do this? By drawing a triangle in a semicircle, without specifying the length of the diameter of the circle or the orientation of the two sides of the triangle in the semicircle, we are, in effect, dealing with the "general case"—the case of every single triangle inscribed in a semicircle.

Let's start by drawing a triangle in a semicircle as stipulated:

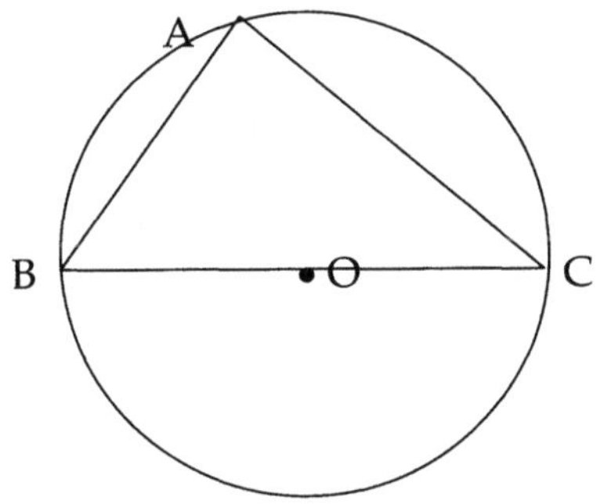

Triangle ABC is a general triangle, because we did not make it to any specification. Indeed, we could have drawn it with A touching any part of the circumference in the semicircle.

Let's start by joining the vertex A of triangle ABC to the center of the circle O, producing the straight line AO:

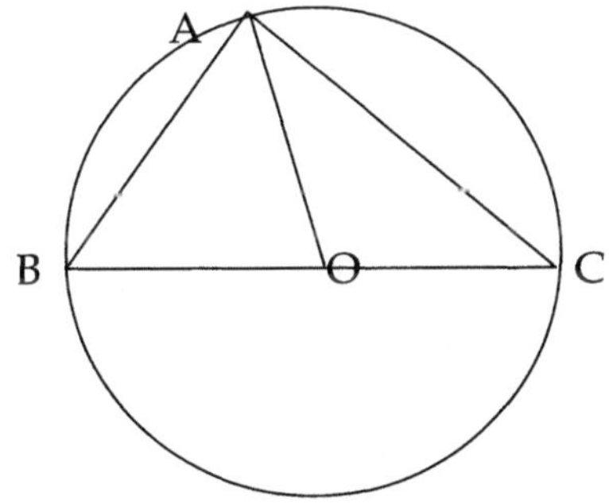

We point out that AO is a radius of the circle. So too are OB and OC. These lines are thus all equal to each other. In the semicircle there are now two isosceles triangles—AOB and AOC. The angles opposite the equal sides of an isosceles triangle are equal. The equal angles in triangle AOB can be labeled x, and those in triangle AOC can be labeled y:

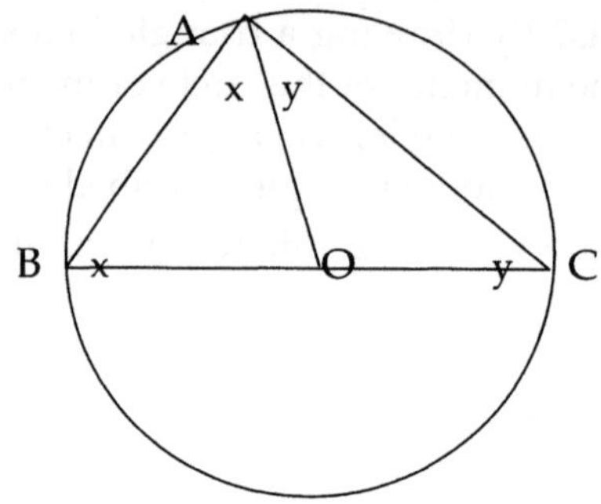

Now, let's consider the original triangle ABC. The sum of the angles in a triangle is equal to 180^0. Therefore, the sum of the angles in triangle ABC can be represented with the following equation:

$$x + x + y + y = 180^0$$

Simplifying the left side of the equation, we get:

$$2x + 2y = 180^0$$

Dividing both sides by 2, the equation simplifies further to:

$$x + y = 90^0$$

Now, $x + y$ is the total number of degrees in the angle at the vertex A, namely $\angle BAC$. Since we have just proven that $x + y$ is equal to 90^0, we can conclude that $\angle BAC$ is equal to 90^0. This is an example of a deductive solution. It involves applying previous knowledge to a problem.

As discussed in the previous chapter, induction and deduction are two main logical strategies used in PS, especially closed PS. A third major one, called abduction, is the one however that is usually entailed for the solution of open problems, such as the Nine-Dot one (problem [5–24]). As mentioned, most mathematical discoveries come about not by induction or deduction, but by abduction. The former two are useful as conceptual strategies for consolidating the discoveries made by abductive thinking: that is, they allow for mathematical discoveries to be organized into systems of mathematical knowledge. Once an insight is attained, it becomes useful to routinize it, so that a host of related problems can be solved as a matter of course, with little time-consuming mental effort. Such routinization is a memory-preserving and time-saving strategy. It is the rationale behind algorithms—routinized procedures for solving problems that would otherwise require insight thinking to be used over and over again. In effect, all three forms of logic are required in PS and math generally.

Set Theory

As discussed briefly in chapter 1, mathematics and semiotics have had a parallel history. Devising efficient ways to represent the innate instinct for counting is the sum and substance of mathematics; studying those ways is the sum and substance of semiotics. This overlap is the reason why a semiotic framework for understanding problems and for teaching PS has been proposed in this book. Discovery in math has always been guided by the representations used or invented by mathematicians. In other words, discovery in math is, typically, a serendipitous consequence of semiosis (the use of signs to understand things). Analogously, discovering how to solve problems successfully entails knowing how to represent problems effectively. The relevant research discussed in this book has shown that students already have this ability, but are not always capable of using it for formal learning tasks and for carrying out PS. The goal of semiotic pedagogy is to tap into this ability, bringing it out in the open, so to speak, by teaching it explicitly.

In many ways the key concept in both math and semiotics is that of code. Codes are collections of signs that form a system through a basic structural feature—opposition. This feature allows people to select and combine the signs in specific ways to construct messages, carry out actions, enact rituals, and so on, in meaningful ways. Stories, fashion styles, musical compositions, etc. are all code-based productions. In math, arithmetic, algebra, and the other areas are all codes, as we have seen, each with its particular oppositions that allows us to carry out counting at various levels of abstraction. In other words, mathematical codes are systems for grouping elements with common properties together—integers, fractions, polygons, linear equations, etc. Mathematicians do not refer to these as codes, however, but rather as *sets*. It is instructive to note that the theoretical study of mathematical structure now comes under the rubric of *set theory*, a framework which emerged in the nineteenth century when the ideas of Augustus De Morgan (1806–1871) and Georg Cantor (1845–1918) laid the groundwork for its development. Elementary set theory is the last topic to be treated in this book, because it brings out the fact that mathematics is a code-based phenomenon. Suffice it to say that I will only scratch the surface here.

Capital letters are used to name sets and lower-case ones to name the members of sets. For example, the letter S might stand for the set of all girls with straight hair. The letters m, s, and r might then be used to represent three members of this set—Martha, Sarah, and Ruth. So, the set of all girls with straight hair is: $S = \{m, s, r, \ldots\}$. The symbol $\in$ is used to indicate that a member belongs to a certain set. For example, $s \in S$ is read: "s is a member of S."

There are several main types of sets:

- *Universal sets*, represented with U, consist of all members being considered at any one time. For example, the set of all the positive integers is U = {0, 1, 2, 3, 4, 5, 6, 7, 8, 9, …}.

- This is an *infinite set* because it has an endless number of members. The set of integers is also infinite.

- A *finite set*, on the other hand, has a definite number of members. One such set is the set of natural single digits including zero: {0, 1, 2, 3, 4, 5, 6, 7, 8, 9}.

- An *empty set*, also called a *null set*, has no members. The symbol ø is used to show this set: {ø}. An example of an empty set is all odd numbers that end in 2—there is no such set of course.

- A *single element set* contains only one member. For example, the set of all primes less than or equal to 2 contains only 2.

- *Equivalent* or *equal sets* have the same number of members. For instance, the set of even numbers under ten, {0, 2, 4, 6, 8}, is equal to the set of odd numbers under ten, {1, 3, 5, 7, 9}.

- *Overlapping sets* have some members in common. If the set of last year's class math stars is M_1 = {Alex, Sarah, Betty} and the set of this year's stars is M_2 = {Alex, Sarah, Tom}, sets M_1 and M_2 overlap because Alex and Sarah belong to both sets. This relation between sets is usually shown with intersecting circles in which the common members are included in the area of overlap: (a = Alex, s = Sarah, b = Betty, t = Tom):

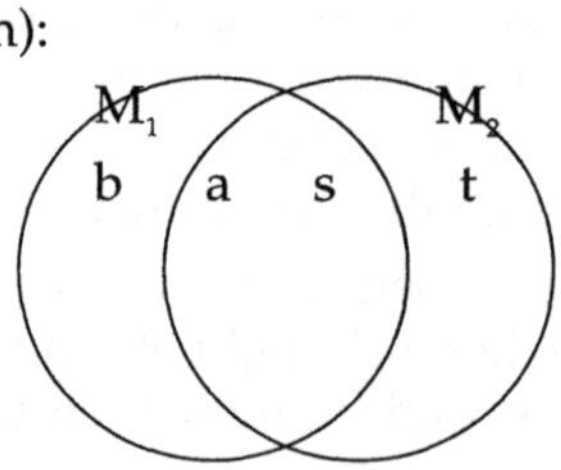

- *Disjoint sets* have no members in common. The set of even numbers and the set of odd numbers are disjoint because they do not have any elements in common.

- *Subsets* are sets contained within other sets. For example, the set of even numbers, E = {0, 2, 4, 6, 8, ...}, is a subset of the set of all integers, I = {0, 1, 2, 3, 4, 5, 6, ...}. This is shown with $E \subset I$. The symbol "$\subset$" is read: "is included in."

Set theory, like code theory in semiotics, makes it possible to envision what would otherwise be seen as disparate elements as having structural equivalence. Incidentally, diagrams such as the overlapping circles above are called *Venn diagrams,* named after British logician John Venn (1834–1923), who was the first to use them. These are concretization devices, as they have been called in this book, and thus can be applied to the solution of problems of various kinds. Here's an example:

> [PROBLEM 6–7]
> In a school, each single student is involved in hockey or soccer or in both. Can you figure out how many students there are in the school if we are told that 180 students are involved in hockey, 70 percent in soccer, and 10 percent in both hockey and soccer?

The students involved in the two sports can be shown with Venn diagrams as follows:

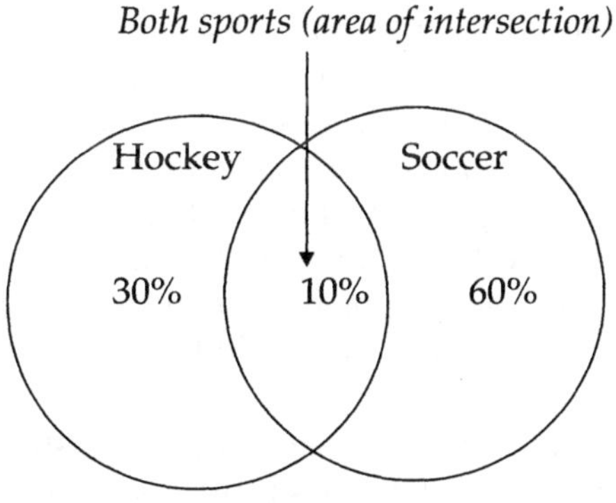

This diagrammatic representation of the problem shows that those involved in both sports (the area of intersection) is 10%. Therefore, the right-hand circle, which represents those involved in soccer, is completed with 60% in the section of the circle representing students enrolled just in soccer (since 70% were involved in soccer). Since 100% of the students were involved in one or the other sport, this leaves 30% for those involved just in hockey (100% - 70%) in the sector shown in the left-hand circle. There are 180 students in hockey. This equals 40% of the total number of students (the entire left-hand circle). So, the answer is that 180 is 40% of the total number of students in the school. Thus the answer is 450 (40% of 450 = 180).

The theory of sets, like any semiotic construct, has led to discoveries, many of them fundamental to the contemporary study of mathematics. It is interesting and important to bring this out in the classroom as a contextualization technique. For example, it can be pointed out that set theory may have actually started in the sixteenth century when the great Italian scientist Galileo Galilei (1564–1642) noticed a pattern—if the counting numbers, {1, 2, 3, 4, …}, are compared, one-by-one, to the numbers in one of its subsets in sequence, such as the square numbers {1, 4, 9, 16, …}, something astounding happens:

1	2	3	4	5	6	7	8	9	10	11	12	13	14	…
↕	↕	↕	↕	↕	↕	↕	↕	↕	↕	↕	↕	↕	↕	…
1	4	9	16	25	36	49	64	81	100	121	144	169	196	…

No matter how far one continues this layout of numbers, there will never be a gap between the top line and the bottom one. This suggested to Galileo that the "number" of elements in the set of all positive integers and the "number" of those in one of its proper subsets is the same. If one stops to think about it, this is indeed an astonishing discovery. To this day, students have a hard time grasping its implications. A study by Wheeler (1987) showed, in fact, that while many students will agree that two infinite sets such as the set of counting numbers and the set of even numbers are equivalent, the same students will argue that the set of counting numbers is larger due to its inclusion of the odd numbers. As Clark (2007: 68) aptly puts it: "We are so accustomed to thinking of finite collections that our intuitions become disturbed when we first consider infinite sets like the set of positive integers."

It is instructive to note that such seemingly "disturbing" results, as Clark characterizes them, derived from simply using a diagrammatic line representation of the two sets of numbers and putting them in a one-to-one relation. In semiotic terms, it was the structure of the layout that suggested the pairing technique. This no doubt inspired Georg Cantor's (1845–1918) truly incredible demonstrations about infinite sets around 1870. Like Galileo he showed that the set of counting numbers, also called *cardinal* numbers, can be matched against any of its subsets, such as the even numbers:

1	2	3	4	5	6	7	8	9	10	11	12	13	14	...
↕	↕	↕	↕	↕	↕	↕	↕	↕	↕	↕	↕	↕	↕	...
2	4	6	8	9	10	11	12	13	14	15	16	17	18	...

Cantor pointed out that such sets have the same "cardinality," or same number of elements. He represented this cardinality with the Hebrew letter $\aleph_0$ (aleph null):

$$\aleph_0 = \{1, 2, 3, 4, \ldots\}$$

Any infinite set that can be put in a one-to-one correspondence with $\aleph_0$ has the same cardinality (size). He called $\aleph_0$ a *transfinite* number, and then discovered something else that is even more mind-boggling—the transfinite numbers themselves constitute an infinite set, $\{\aleph_0, \aleph_1, \aleph_2, \aleph_3, \ldots\}$, each one describing sets with different cardinalities. However, not all number sets have the same cardinality. The set of real numbers has a greater cardinality than $\aleph_0$.

Incidentally, teaching the notion of sets through historical annotations makes it possible to contextualize appropriate problems effectively. Following are two such problems that can be so devised:

[PROBLEM 6–8]

What type of set is each of the following?

(1) $\{0\}$

(2) $A = \{2, 4, 6, 8, \ldots\}$, $B = \{2n\}$, $n = \{0, 1, 2, 3, \ldots\}$

(3) $A = \{0, 2, 4, 6, 8\}$, $B = \{0, 1, 3, 5, 7, 9\}$

(4) $A = \{1, 2, 3, 4, 5, 6, 7, 8, 9\}$, $B = \{2, 4, 6, 8\}$

[PROBLEM 6–9]

Using Cantor's simple method, show that the following sets have the same cardinality, or the same number of members in them:

(1) the odd numbers

(2) successive powers of 10

(3) cube numbers (numbers that are raised to the power of three)

(4) all numbers ending in 5

Overview and Reflections

The teaching principles, techniques, and ideas discussed and illustrated in this book constitute the core of what I have called semiotic pedagogy. As such, they do not comprise something new in math education. They have always been the sum and substance of effective math teaching. But such teaching has rarely (if ever) been fashioned into a consistent framework for conducting classroom teaching. So, the techniques have tended to be used in sporadic ways, disconnected from any overall framework for building math competence. It is the framework itself that, in my view, will improve learning outcomes considerably and, potentially, build such competence. The implicit claim I have interweaved throughout this book is that an appropriate pedagogical framework for teaching PS should activate a flow of learning that goes from an experiential mode of understanding to an abstract one. The former mode is best set in motion with the use of iconic techniques (diagrams, schemata, annotation, contextualization, etc.), and the latter mode with more formal techniques (analysis, generalization, etc.). For the sake of convenience, I will call this the *natural semiotic learning flow* (NSLF).

Several studies substantiating, or at least supporting, the NSFL principle have been described in previous chapters. The gist of the studies is that students taught according to this principle tend to do better on standardized math tests than do comparable groups. The principle is, however, not an all-embracing theory of learning. It simply offers a semiotic perspective on how to view typical learning behaviors. As such, it does not exclude using other pedagogical techniques. The NSFL posits that novel learning is grounded in the experiential realm of the senses. This initial stage of knowing can be called a *cognizing* stage, while the stage when the sensory units of knowing are transformed into more symbolic formal modes can be called the *recognizing* stage. The NSFL principle can thus be seen to be a tenable general principle of learning applicable at all stages of education, from elementary school to college. It claims, essentially, that novel learning tasks (*cognizing events*) are apt to be more comprehensible if they are presented iconically, whereas practice of already-acquired knowledge and skill *(recognizing events)* require a more conventionalized, symbolic form of presentation and practice. The key to successful learning lies, therefore, in determining at what point the student's *cognizing mode* is ready to become a *recognizing mode*. As Pamela Liebeck (1984: 15–16) has aptly observed, this view of learning actually follows the stages of cognitive development in childhood documented by psychology—a series of stages that have to be repeated in adolescence and adulthood when concepts are new, unfamiliar, or acquired previously in an erroneous manner. Incidentally, this view of learning is also consistent with the model posited by well-known educator Zoltan Dienes

(1964). For Dienes, a direct sensory engagement with learning materials is the crucial first stage in effective learning (cognizing mode). After this stage, Dienes suggests, a secondary abstract form is required (recognizing mode).

I should mention, as one of my final reflections, that not everyone working in semiotics accepts the perspective of learning that I have delineated here. Let me just say that this lack of consensus is inevitable. As in any other scientific domain, semiotics too has its debates, its controversies, its different theories. Semiotics is a dynamic, vibrant, ever-changing science. If nothing else, the NSFL principle is at least worth testing out in future educational research.

The techniques of concretization, annotation, contextualization, generalization, and exemplification, for example, are drafted directly from this principle. Their objective is to get the student to grasp the structure of mathematical concepts and the correlative problems constructed on that structure through iconicity. The corpus of psychological studies in math education has confirmed this aspect of the NSLF principle. A pivotal study by Yancey, Thompson, and Yancey (1989), for instance, has shown rather conclusively that training students how to use pictures (diagrams, charts, etc.) to solve problems results in improved performance. As Musser, Burger, and Peterson (2006: 20) have aptly put it: "All students should represent, analyze, and generalize a variety of patterns with tables, graphs, words, and, when possible, symbolic rules." But the NSFL also stresses that formalization must not be excluded from the pedagogical process, since it is the point of arrival in the learning flow. The task of semiotic pedagogy is to create learning conditions that activate this flow. If it is not activated, or if only the experiential cognizing mode is stressed, effective global learning rarely occurs. A significant study by Ambrose (2002) suggests, in fact, that arithmetic students who are taught appropriately with concrete strategies, but not allowed to develop their own abstract representational grasp of the arithmetical code, are less likely to develop arithmetical fluency.

A simple sequential model of what semiotic pedagogy entails in terms of the NSLF is the following one. This summarizes the notions discussed throughout this book:

During the Cognizing Mode Stage

PROBLEMS	→	Should be devised in both open and closed fashion to allow students to grasp the nature of PS.
↓		
CONCRETIZATION	→	Students should be taught how to represent problem content iconically (with dia-

grams, schemata, charts, figures, etc.).

↓

ANNOTATION → Appropriate commentary on the steps used in the PS process should be used. Such commentary is an indexical technique.

↓

CONTEXTUALIZATION → Real-life examples or historical anecdotes should always be used to reinforce the learning process because these provide appropriate contexts in which learning unfolds.

↓

EXEMPLIFICATION → Closed and open problem texts can then be devised to exemplify the structural properties of given problems.

↓

EXPLORATION → After basic structures have been acquired, heuristic problems should be used so that students can explore the structural principles presented and expand upon them on their own.

During the Recognizing Mode Stage

FORMALIZATION → Students should be shown how to generalize from specific problems using symbols, formulas, etc.

↓

ALGORITHMICIZATION → Algorithms for routinizing PS should be introduced and studied.

↓

LOGICAL STRATEGIES → A discussion of the logical strategies suggested by problem representations should be discussed formally.

↓

DECODING → Students should be introduced to basic linguistic analysis and CMT to help them set up equations.

In line with the basic supposition of semiotic pedagogy that one cannot assume that students already have the know-how required to represent problems, a significant study conducted by Clement, Lochhead, and Monk

(1981) found, in fact, that it must be taught explicitly: "the modeling process is far more complex than is generally imagined because rather than being an immediate aid to learning mathematics (1981: 287). As Dolan and Williamson (1983: ix) have aptly observed, to teach PS successfully, "we must do more than simply assign problems to students."

As mentioned in the preface to this book, the motivation for writing it came initially from an e-mail sent to me by a student who took an experimental math course from me. With this book I have attempted to answer her behest, by trying to shed some light on how semiotics can be incorporated generally into math education. Much work remains to be done on expanding the notions discussed here and in testing their overall validity. And I must emphasize one last time that the semiotic approach to math education is not the "magic key" that will guarantee successful learning in all situations. Like good music teachers, experienced math teachers know all too well that each student is different, that each learning task presents its own kinds of difficulties, and, thus, that *patience* is the operative word in all situations and contexts of learning.

GLOSSARY

This glossary contains only those terms that are related to the pedagogical ideas discussed in this book. Common mathematical notions (number, prime, equations, etc.) are not included here.

abduction	logical thinking based on informed hunches and inferences
analysis	interpreting the content of a story problem through diagrammatic representation
annotation	any pedagogical strategy serving as a guide, or identifying in concrete terms what the stages and features of a solution are.
closed problem	type of problem that presents all information needed for its solution in an unambiguous fashion
code	any system of signs having the same properties and governed by the same rules of structure (for example the alphabet code, the integer code, etc.)
conceptual metaphor	thought formula that results from thinking of abstract notions in terms of concrete experiences (for example, thinking of love as an experience akin to gustation leads to the conceptual metaphor *love is sweet*)
concretization	any pedagogical strategy that represent the information contained in a problem concrete and thus more easily comprehensible (diagrams, charts, etc.)
contextualization	the pedagogical technique of providing real-world applications, analogues, historical anecdotes, complementary information, etc. to illustrate some notion
deduction	showing that something is logically the way it is by applying previously-established facts to the demonstration
diagram	any drawing, graph, schema, figure, etc. showing the relevant information in a problem in outline form
exemplification	process of constructing or adapting problems and activities for students that will allow them to utilize and apply the knowledge they have gained
exploration	a strategy that allows a learner to use new information or skills in an exploratory fashion to see what they yield
generalization	strategy of representing something in some way (usually with a formula) so as to reveal its structure
heuristics	using problems as guides to exploration of mathematical structure, that is, as guides for learning

icon a sign that stands for something through resemblance (a diagram, a schema, etc.)

iconicity process of using *iconic* techniques in teaching (diagrams, schemas, etc.); process involving any form of resemblance or simulation

index a sign that indicates something in relation to something else (road arrows, for example, show where something is located relative to the observer or the arrow)

indexicality process of using *indexical* techniques in teaching (arrows, annotations, etc.); process involving any form of relation

induction generalizing, proving, or solving a problem on the basis of given facts and extracting a pattern from them

marked form form in a binary *opposition* that is the derivative of the other form (called *marked* or default) or conceptualized in reference to it

markedness theory that forms relate to each other in terms of a binary *opposition* whereby one is marked (exceptional) and the other unmarked (typical)

model any representamen that displays something in a concrete and easily understandable way (for example, the number line can be used to display how the arithmetical operations are performed to reflect counting processes

morpheme minimal unit standing for something on its own (for example, the *im-* in *improper* is a prefix morpheme)

morphology study or analysis of the form (or structural make-up) of signs (for example, the word *illogical* is made up morphologically of a prefix (*il-*), a freely occurring form (*logic*), and a suffix (*-al*)

naming giving names to mathematical symbols, structures, forms, and layouts a mnemonic annotative strategy (for example, by naming the end result of adding numbers a *sum* we can reinforce what the result stands for conceptually)

open problem type of problem that does not present all the information needed for its solution, requiring a more creative solution strategy (also known as *puzzle*)

opposition theory that forms are differentiated psychologically through some minimal contrast in their make-up or their meaning

paradigmatic feature of codes (and the signs in them) that keeps signs distinct and recognizable

Principle of Least Effort theory that there is a tendency towards least effort in human representation and communication systems; that is to say, over time, people tend to economize the struc-

	ture of common forms (words, sentence length, mathematical symbolism, mathematical operation procedures)
problem	a type of question, consisting of information or a set of conditions designed to elicit a particular answer
puzzle	an *open problem*
representamen	term used for sign-form, that is, the form used to represent the elements of a problem
representation	process of representing (literally "presenting again" in some other form) the elements of a problem
schema	a type of diagram that mirrors the actual shape or form of something
semiosis	the innate ability to produce and comprehend signs
semiotics	the science studying signs, their meanings, and their evolution in both the human species and in individual cultures
sign	anything that stands for something other than itself (for example, the letter x in mathematics does not stand for that letter, but for some unknown quantity or number)
signified	the part of the sign to which the sign refers; also called the referent or the concept (for example, the word *tree* refers to an arboreal plant, which is the signified)
signifier	the physical part of the sign (its actual physical structure, such as the sounds that make up the word *tree*)
structure	feature of all signs showing how they are constructed systematically and how they relate to each other
symbol	any sign that stands for something in an abstract or general way (for example 2n stands for any even number, no matter what value n takes on)
syntagmatic	feature of codes (and the signs in them) that specifies how the signs are combined or organized
symbolicity	process of using *symbolic* techniques in teaching (formulas, rules, etc.); process involving any form of abstract or general representation
transference	translating the language of a story problem into appropriate algebraic language
unmarked form	form in a binary *opposition* that is the typical representation of the class or category for which the opposition stands
Zipf's Law	notion that the length of a form correlates negatively with its frequency in a text; that is, the higher its frequency the shorter it tends to be

REFERENCES

Adam, J. A. (2004). *Mathematics in Nature: Modeling Patterns in the Natural World.* Princeton: Princeton University Press.

Adams, T. L. (1993). *The Effects of Graphing Calculators and a Model for Conceptual Change on Community College Students' Concept of Function.* University of Florida Dissertation.

Ambrose, C. (2002). Are We Overemphasizing Manipulatives in the Primary Grades to the Detriment of Girls? *Teaching Children Mathematics* 9: 16–21.

Anderson, M., Sáenz-Ludlow, A., and Cifarelli, V. (2000). Musement in Mathematical Manipulation. In: A. Gimate-Welsh (ed.), *Ensayos Semióticos*, pp. 663–676. Mexico: Porrúa.

Anderson, M., Sáenz-Ludlow, A., and Cifarelli, V. (eds.) (2003). *Educational Perspectives on Mathematics as Semiosis: From Thinking to Interpreting to Knowing*. Ottawa: Legas Press.

Antenos-Conforti, E., Barbeau, E., and Danesi, M. (1998). *Problem-Solving in Mathematics: A Semiotic Perspective with Educational Implications*. Welland: Soleil (Toronto: Monograph Series of the Toronto Semiotic Circle, 20).

Bakst, A. (1967). *Mathematics: Its Magic and Mastery*. Princeton: Van Nostrand.

Barbeau, E. (1992). What Is a Problem? In: C. Cicogna, M. Danesi, and A. Mollica (eds.), *Problem-Solving in Second-Language Teaching*, pp. 1–12. Welland: Soleil.

Beckmann, P. (1981). *A History of* . New York: St. Martin's.

Bentley, P. J. (2008). *The Book of Numbers: The Secret of Numbers and How They Changed the World*. Buffalo: Firefly.

Blatner, D. (1997). *The Joy of Pi*. Harmondsworth: Penguin.

Boers, F. (2000). Metaphor Awareness and Vocabulary Retention. *Applied Linguistics* 21: 553–571.

Brainerd, B. (1971). *Introduction to the Mathematics of Language Study*. New York: American Elsevier.

Bronowski, J. (1973). *The Ascent of Man*. Boston: Little, Brown, and Co.

Brooke, M. (1969). *150 Puzzles in Crypt-Arithmetic*. New York: Dover.

Bruner, J. (1968). *Towards a Theory of Instruction*. New York: Norton.

Bruner, J. et al. (1967). *Studies in Cognitive Growth*. New York: John Wiley.

Cameron, L. (2003). *Metaphor in Educational Discourse*. London: Continuum.

Cameron, L. and Low, G. (eds.) (1999). *Researching and Applying Metaphor*. Cambridge: Cambridge University Press.

Cathcart, W. G., Pothier, Y. M., and Vance, J. H. (1994). *Learning Mathematics in Elementary and Middle Schools*. Scarborough: Allyn and Bacon.

Chase, A. (1979). *The Rhind Mathematical Papyrus: Free Translation and Commentary with Selected Photographs, Transcriptions, Transliterations and Literal Translations*. Reston, VA: National Council of Teachers of Mathematics.

Churchman, C. W. (1970). The Role of Weltanschauung in Problem Solving and Inquiry. In: R. B. Banerji and M. D. Mesarovic (eds.), *Theoretical Approaches to Non-Numerical Problem Solving*, pp. 141–51. Berlin: Springer.

Clark, K. M. and Cunningham, D. J. (2006). Metaphors We Teach by: An Embodied Cognitive Analysis of *No Child Left Behind*. *Semiotica* 161: 265–289.

Clark, M. (2007). *Paradoxes from A to Z*. London: Routledge.

Clarke, D. S. (1987). *Principles of Semiotic.* London: Routledge and Kegan.

Clement, J., Lochhead, J., and Monk, G. S. (1981). Translation Difficulties in Learning Mathematics. *American Mathematical Monthly* 88: 286–290.

Cole, K. C. (1984). *Sympathetic Vibrations*. New York: Bantam.

Crawford, J. T., Dean J. E., and Jackson W. A. (1954). *A New Algebra for High Schools*. Toronto: Macmillan.

Croft, W. and Cruse, D. A. (2004). *Cognitive Linguistics.* Cambridge: Cambridge University Press.

Cunningham, D. (1987a). Semiotics and Education: An Instance of the "New" Paradigm. *American Journal of Semiotics* 5: 195–199.

Cunningham, D. (1987b). Outline of an Education Semiotic. *American Journal of Semiotics* 5: 201–216.

Cunningham, D. (1992). Beyond Educational Psychology: Steps toward an Educational Semiotic. *Educational Psychology Review* 4: 165–194.

Cunningham, D. (ed.) (2007). *Semiotics and Education*. Special Issue of *Semiotica* (164). Berlin: Mouton de Gruyter.

Danesi, M. (1982). Alphametics in Different Languages. *Ontario Mathematics Gazette* 21: 27–30.

Danesi, M. (1987). Formal Mother-Tongue Training and the Learning of Mathematics in Elementary School: An Observational Note on the Brussels *Foyer Project*. *Scientia Paedogogica Experimentalis* 24: 313–320.

Danesi, M. (1994). Recent Research on Metaphor and the Teaching of Italian. *Italica* 71: 454–464.

Danesi, M. (1995). Learning and Teaching Languages: The Role of Conceptual Fluency. *International Journal of Applied Linguistics* 5: 3–20.

Danesi, M. (1997). *Increase Your Puzzle IQ*. New York: John Wiley.

Danesi, M. (2002). *The Puzzle Instinct: The Meaning of Puzzles in Human Life*. Bloomington: Indiana University Press.

Danesi, M. (2003). Conceptual Metaphor Theory and the Teaching of Mathematics: Findings of a Pilot Project. *Semiotica* 145: 71–83.

Danesi, M. (2006a). *A "User-Friendly" Course in Basic Mathematics.* Ottawa: Legas Press.

Danesi, M. (2006b). Alphabets and The Principle of Least Effort. *Studies in Communication Sciences* 6: 47–62.

Danesi, M. (2007a). A Conceptual Metaphor Framework for the Teaching Mathematics. *Studies in Philosophy and Education* 26: 225–236

Danesi, M. (2007b). *The Quest for Meaning: A Guide to Semiotic Practice and Theory*. Toronto: University of Toronto Press.

Daniels, H. (ed.) (1996). *An Introduction to Vygotsky.* London: Routledge.

Davydov, V. V. and Radzikhovskii, L. A. (1985). Vygotsky's Theory and the Activity Oriented Approach in Psychology. In: J. V. Wertsch (ed.), *Culture, Communication and Cognition: Vygotskian Perspectives*, pp. 59–69. Cambridge: Cambridge University Press.

Deely, J. (1982). *Introducing Semiotics.* Bloomington: Indiana University Press.

Deely, J. (1990). *Basics of Semiotics.* Bloomington: Indiana University Press.

Deely, J. (2001). *Four Ages of Understanding: The First Postmodern Survey of Philosophy from Ancient Times to the Turn of the Twentieth Century.* Toronto: University of Toronto Press.

Dehaene, S. (1997). *The Number Sense: How the Mind Creates Mathematics.* Oxford: Oxford University Press.

Devlin, K. (1998). *The Language of Mathematics: Making the Invisible Visible.* New York: W. H. Freeman.

Devlin, K. (2000). *The Math Gene: How Mathematical Thinking Evolved and Why Numbers Are Like Gossip.* New York: Basic Books.

Devlin, K. (2005). *The Math Instinct.* New York: Thunder's Mouth Press.

Dienes, Z. (1964). *An Experimental Study of Mathematics Learning.* London: Hutchison.

Dirven, R. and Verspoor, M. (1998). *Cognitive Exploration of Language and Linguistics.* Amsterdam: John Benjamins.

Dolan, D. T. and Williamson, J. (1983). *Teaching Problem-Solving Strategies.* Menlo Park: Addison-Wesley.

Eccles, J. and Jacobs, J. (1986). Social Forces Shape Maths Attitudes and Performance. *Journal of Women in Culture and Society* 11: 367–380.

Eco, U. (1976). *A Theory of Semiotics.* Bloomington: Indiana University Press.

Eco, U. (1979). *The Role of the Reader. Explorations in the Semiotics of Texts.* Bloomington: Indiana University Press.

Eco, U. (1992) *Interpretation and Overinterpretation.* Cambridge: Cambridge University Press.

Eco, U. (1998). *Serendipities: Language and Lunacy,* translated by William Weaver. New York: Columbia University Press.

English, L. D. (ed.) (1997). *Mathematical Reasoning: Analogies, Metaphors, and Images.* Mahwah, N.J.: Lawrence Erlbaum Associates.

Eymard, P., Lafon, J-P, Wilson, S. S., and Lafon, J. P. (2004). *The Number Pi.* New York: American Mathematical Society.

Fauconnier, G. and Turner, M. (2002). *The Way We Think: Conceptual Blending and the Mind's Hidden Complexities.* New York: Basic.

Fixx, J. F. (1983). *Solve It!* London: Muller.

Friedman, Thomas L. (2007). *The World Is Flat: A Brief History of the Twenty-First Century.* New York: Picador.

Gardner, M. (1971). *Sixth Book of Mathematical Games from Scientific American.* New York: Charles Scribner's Sons

Geeraerts, D. (ed.) (2006). *Cognitive Linguistics.* Berlin: Mouton de Gruyter.

Gibbs, R. W. (1994). *The Poetics of Mind: Figurative Thought, Language, and Understanding*. Cambridge: Cambridge University Press.

Gillings, R. J. (1961). Think-of-a-Number: Problems 28 and 29 of the Rhind Mathematical Papyrus. *The Mathematics Teacher* 54: 97–102.

Gillings, R. J. (1962). Problems 1 to 6 of the Rhind Mathematical Papyrus. *The Mathematics Teacher* 55: 61–65.

Gillings, R. J. (1972). *Mathematics in the Time of the Pharaohs*. Cambridge, Mass.: MIT Press.

Gowan, J. C., Khatena, J. and Torrance, E. P. (eds.) (1981). *Creativity: Its Educational Implications*. Dubuque: Kendall/Hunt.

Graeber, A. O. and Tirosh, D. (1988). Extending Multiplication and Division to Decimals: Insights Forth and Fifth Graders Bring to Task. Paper delivered at the symposium, *Understanding Situations Modeled by Multiplication and Division*, American Educational Research Association, New Orleans.

Greimas, A. J. (1987). *On Meaning: Selected Essays in Semiotic Theory*, trans. by P. Perron and F. Collins. Minneapolis: University of Minnesota Press.

Hallyn, F. (1990). *The Poetic Structure of the World: Copernicus and Kepler*. New York: Zone Books.

Hockett, C. F. (1967). *Language, Mathematics and Linguistics*. The Hague: Mouton.

Holme. R. (2004) *Mind, Metaphor and Language Teaching*. Basingstoke: Palgrave.

Hunter, J. A. H. (1965a). *Fun with Figures*. New York: Dover.

Hunter, J. A. H. (1965b). *Hunter's Math Brain Teasers*. New York: Bantam.

Irwin, K. C. (2001). Using Everyday Knowledge of Decimals to Enhance Understanding. *Journal for Research in Mathematics Education* 32: 399–420.

Jacoby, O. (1962). *Mathematics for Pleasure*. Greenwich, Connecticut: Fawcett.

Johnson, M. (1987). *The Body in the Mind: The Bodily Basis of Meaning, Imagination and Reason*. Chicago: University of Chicago Press.

Jones, R. (1982). *Physics as Metaphor*. New York: New American Library.

Kasner, E. and Newman, J. (1940). *Mathematics and the Imagination*. New York: Simon and Schuster.

Kecskes, I. (2000). Conceptual Fluency and the Use of Situation-Bound Utterances in L2. *Links and Letters* 7: 143–158.

Kehle, P. (1999). Shifting our Focus from Ends to Means: Mathematical Reasoning. *Journal for Research in Mathematics Education* 30: 468–474.

Klemme, S. L. (1981). References of Speech Acts as Characteristics of Mathematical Classroom Conversation. *Educational Studies in Mathematics* 12: 43–58.

Knowles, M. and Moon, R. (2006). *Introducing Metaphor*. London: Routledge.

Kövecses, Z. and Szabo, P. (1996). Idioms: A View from Cognitive Semantics. *Applied Linguistics* 17: 326–355.

Kövecses. Z. (2002). *Metaphor: A Practical Introduction*. Oxford: Oxford University Press

Lakoff, G. (1987). *Women, Fire, and Dangerous Things: What Categories Reveal about the Mind*. Chicago: University of Chicago Press.

Lakoff, G. and Johnson, M. (1980). *Metaphors We Live By*. Chicago: Chicago University Press.

Lakoff, G. and Johnson, M. (1999). *Philosophy in the Flesh: The Embodied Mind and its Challenge to Western Thought*. New York: Basic Books.

Lakoff, G. and Núñez, R. E. (2000). *Where Mathematics Comes From: How the Embodied Mind Brings Mathematics into Being*. New York: Basic Books.

Langacker, R. W. (1987). *Foundations of Cognitive Grammar*. Stanford: Stanford University Press.

Langacker, R. W. (1990). *Concept, Image, and Symbol: The cognitive basis of grammar*. Berlin: Mouton de Gruyter.

Langacker, R. W. (1999). *Grammar and Conceptualization*. Berlin: Mouton de Gruyter.

Lave, J. (1988). *Cognition in Practice: Mind, Mathematics and Culture in Everyday Life*. Cambridge: Cambridge University Press.

Leezenberg, M. (2001). *Contexts of Metaphor*. Amsterdam: Elsevier.

Lemke, J. L. (1987). Social Semiotics and Science Education. *American Journal of Semiotics* 5: 211–232.

Lesgold, A. (1988). Problem Solving. In: R. J. Sternberg and E. E. Smith (eds.), *The Psychology of Human Thought*, pp. 188–213. Cambridge: Cambridge University Press.

Lévi-Strauss, C. (1958). *Structural Anthropology*. New York: Basic Books.

Liebeck, P. (1984). *How Children Learn Mathematics: A Guide for Parents and teachers*. Harmondsworth: Penguin.

Loyd, S. (1914). *Cyclopedia of Tricks and Puzzles*. New York: Dover.

Loyd, S. (1959–1960). *Mathematical Puzzles of Sam Loyd*, 2 volumes, compiled by M. Gardner. New York: Dover.

MacGregor, M and Price, E. (1999). An Exploration of Aspects of Language Proficiency and Algebra Learning. *Journal for Research in Mathematics Education* 30: 449–467.

MacNamara, O. (1996). Mathematics and the Sign. *Proceedings of PME* 20: 369–378.

Maor, E. (1994). *e: The Story of a Number*. Princeton: Princeton University Press.

Maor, E. (2007). *The Pythagorean Theorem: A 4,000-Year History*. Princeton: Princeton University Press.

McKellar, D. (2007). *Math Doesn't Suck: How to Survive Middle School Math*. New York: Hudson Street Press.

Merton, R. K. and Barber, E. (2003). *The Travels and Adventures of Serendipity: A Study in Sociological Semantics and the Sociology of Science*. Princeton: Princeton University Press.

Moore, G. (1951). *Algebra*. New York: Barnes and Noble.

Morris, C. (1938). *Foundations of the Theory of Signs*. Chicago: University of Chicago Press.

Morris, C. (1946). *Signs, Language and Behavior*. Englewood Cliffs, N. J.: Prentice-Hall.

Mott-Smith, G. (1954). *Mathematical Puzzles*. New York: Dover.

Mulligan, J. T. and Mitchelmore, M. C. (1997). Young Children's Intuitive Models of Multiplication and Division. *Journal for Research in Mathematics Education* 28: 309–330.

Munzert, A. W. (1991). *Test Your IQ*. New York: Prentice-Hall.

Musser, G. L., Burger, W. F., and Peterson, B. E. (2006). *Mathematics for Elementary Teachers: A Contemporary Approach*. Hoboken: John Wiley.

Norman, D. A. (1993). *Things that Make Us Smart: Defending Human Attributes in the Age of the Machine*. Reading, Mass.: Addison-Wesley.

Nöth, W. (1990). *Handbook of Semiotics*. Bloomington: Indiana University Press.

Olivastro, D. (1993). *Ancient Puzzles: Classic Brainteasers and Other Timeless Mathematical Games of the Last 10 Centuries*. New York: Bantam.

Osgood, C. E., Suci, G. J., and Tannenbaum, P. H. (1957). *The Measurement of Meaning*. Urbana: University of Illinois Press.

Peet, T. E. (1923). *The Rhind Papyrus*. Liverpool: University of Liverpool Press.

Peirce, C. S. (1931–1958). *Collected Papers*. Cambridge, Mass.: Harvard University Press.

Perelman, Y. I. (1979). *Algebra Can Be Fun*. Moscow: MIR.

Petrilli, S. and Ponzio, A. (2005). *Semiotics Unbounded: Interpretive Routes through the Open Network of Signs*. Toronto: University of Toronto Press.

Piaget, J. (1952). *The Child's Conception of Number*. London: Routledge and Kegan Paul.

Piaget, J. (1969). *The Child's Conception of the World*. Totowa: Littlefield, Adams and Company.

Pimm, D. (1995). *Symbols and Meanings in School Mathematics*. London: Routledge.

Pollio, H., Barlow, J., Fine, H., and Pollio, M. (1977). *The Poetics of Growth: Figurative Language in Psychology, Psychotherapy, and Education*. Hillsdale, NJ: Lawrence Erlbaum Associates.

Polya, G. (1957). *How to Solve It*. New York: Doubleday.

Posamentier, A. S. (2004). *Pi: A Biography of the World's Most Mysterious Number*. New York: Prometheus.

Posamentier, A. S. and Lehmann, I. (2007). *The (Fabulous) Fibonacci Numbers*. Amherst: Prometheus.

Post, T. R., Wachsmuth, I., Lesh, R., and Behr, M. J. (1985). Order and Equivalence of Rational Numbers: A Cognitive Analysis. *Journal for Research in Mathematics Education* 16: 18–36.

Quinlan, C. (1995). Analyzing Teaching/Learning Strategies for Algebra. In: B. Atweh and S. Flavel (eds.), *Proceedings of the 18th Annual Conference of Mathematics in Australia*, pp. 459–464. Darwin: Northern Territory University.

Radford, L. and Grenier, M. (1996). On Dialectical Relationships between Signs and Algebraic Ideas. *Proceedings of PME* 20: 179–186.

Reed, D. (1994). *Figures of Thought: Mathematics and Mathematical Texts*. London: Routledge.

Rosch, E. (1973a). On the Internal Structure of Perceptual and Semantic Categories. In: T. E. Moore (ed.), *Cognitive Development and Acquisition of Language*, pp. 111–144. New York: Academic.

Rosch, E. (1973b). Natural Categories. *Cognitive Psychology* 4: 328–350.

Rotman, B. (1988). Towards a Semiotics of Mathematics. *Semiotica* 72: 1–35.

Russell, D. S. (1960). *Algebra Problems*. New York: Barnes and Noble.

Salem, L., Testard, F., and Salem, C. (1992). *The Most Beautiful Mathematical Formulas*. New York: John Wiley.

Saussure, F. de. (1916). *Cours de linguistique générale*. Paris: Payot.

Schneider, H. (1965). *Solving Math Word Problems*. Woodland Hills, California: Word/Fraction Math Aid Co.

Schoenfeld, A. (1985). *Mathematical Problem Solving*. New York: Academic.

Sebeok, T. A. (2001). *Signs: An Introduction to Semiotics*, 2nd edition. Toronto: University of Toronto Press.

Sebeok, T. A. and Danesi, M. (2000). *The Forms of Meaning: Modeling Systems Theory and Semiotics*. Berlin: Mouton de Gruyter.

Skemp, R. R. (1971). *The Psychology of Learning Mathematics*. Harmondsworth: Penguin.

Stewart, I. (2001). Foreword to I. Moscovich, *1000 PlayThinks*. New York: Workman Publishing.

Strohmeier, J. and Westbrook, P. (1999). *Divine Harmony: The Life and Teachings of Pythagoras*. Berkeley, CA: Berkeley Hills Books.

Swafford, J. O. and Langrall, C. W. (2000). Grade 6 Students' Preinstructional Use of Equations to Describe and Represent Problem Situations. *Journal for research in Mathematics Education* 31: 89–110.

Trubetzkoy, N. (1968). *Introduction to the Principles of Phonological Description*. The Hague: Martinus Nijhoff.

Valeva, G. (1996). On the Notion of Conceptual Fluency in a Second Language. In: A. Pavlenko and R. Salaberry (eds.), *Papers in Second Language Acquisition and Bilingualism*, pp. 22–38. Ithaca: Cornell University Press.

Van Hiele, P. M. (1984). The Child's Thought and Geometry. In: D. Fuys, D. Geddes, and R. Tischler (eds.), *English Translations of Selected Writings of Dina van Hiele-Geldof and P. M. van Hiele*, pp. 243–252. Brooklyn: Brooklyn College of Education.

Varelas, M. (1989). Semiotic Aspects of Cognitive Development: Illustrations from Early Mathematical Cognition. *Psychological Review* 100: 420–431.

Velasco Sacristán, M. (2005). Metaphor and ESP: Metaphor as a Useful Device for Teaching L2 Business English Learners. *Ibérica* 10: 115–131.

Vygotsky, L. S. (1962). *Thought and Language*. Cambridge, Mass.: MIT Press.

Vygotsky, L. S. (1978). *Mind in Society*. Cambridge, MA: Cambridge University Press.

Wheeler, M. M. (1987). Research into Practice: Children's Understanding of Zero and Infinity. *Arithmetic Teacher* 35: 42–44.

Whitlin, P. and Whitlin, D. (2002). Promoting Communication in the Mathematics Classroom. *Teaching Children Mathematics* 9: 205–211.

Yancey, A., Thompson, C., and Yancey, J. (1989). Children Must Learn to Draw Diagrams. *Arithmetic Teacher* 36: 15–19.

Zipf, G. K. (1935). *Psycho-Biology of Languages*. Boston: Houghton-Mifflin.

Zipf, G. K. (1949). *Human Behavior and the Principle of Least Effort*. Boston: Addison-Wesley.

ONLINE RESOURCES

The sites listed below were operating at the time this book was being written. Some of them may no longer be operating.

Math and Problem-Solving Sites

http://www.seanet.com/~ksbrown

This site contains a number of clear explanations of mathematical topics, ranging from number theory and geometry through probability, statistics, calculus, and the history of math. Many of these are semiotically-oriented, although this is not acknowledged as such. The site is thus useful for constructing problems and for retrieving historical information.

http://forum.swarthmore.edu/dr.math/abt.drmath.html

This site is useful for retrieving information on elementary math concepts. Moreover, teachers can submit questions directly by using the link provided.

http://www2.hawaii.edu/suremath/intro_algebra.html

This site contains many story problems. It can thus be used for modeling the construction of one's own story problems.

http://www.geom.umn.edu

This site gives teachers access to notes, handouts, and other educational materials developed by geometry teachers. It is funded by the National Science Foundation.

Semiotic Sites

http://www.aber.ac.uk/media/Documents/S4B/sem08.html

This is Daniel Chandler's *Semiotics for Beginners* website, which is probably the best online resource for getting more information on semiotics, given its various hyperlinks.

http://carbon.cudenver.edu/~mryder/martin.html

Martin Ryder's website is a storehouse of information and links related to a vast array of topics in semiotics.

www.nist.gov/dads/HTML/zipfslaw.html

This site provides information, works, and materials related to the investigation of Zipf's law.

INDEX